Radioimmunoassays for Insulin, C-Peptide and Proinsulin

Radioimmunoassays for Insulin, C-Peptide and Proinsulin

Lise G. Heding
Novo Research Institute

MTP PRESS LIMITED
a member of the KLUWER ACADEMIC PUBLISHERS GROUP
LANCASTER / BOSTON / THE HAGUE / DORDRECHT

Denne afhandling er i forbindelse med omstående tidligere offentliggjorte afhandlinger af Det lægevidenskabelige Fakultet ved Københavns Universitet antaget til offentligt at forsvares for den medicinske doktorgrad.

København, den 27. oktober 1987

Kjeld Møllgård
dekan

This thesis in conjunction with the previously published papers adjoined hereto has been accepted by the Medical Faculty of the University of Copenhagen for public defence in fulfilment of the requirements for the Degree of Doctor of Medicine.

Copenhagen, October 27, 1987

Kjeld Møllgård
dean

Published in the UK and Europe by
MTP Press Limited
Falcon House
Lancaster, England

ISBN-13: 978-94-011-7096-3   e-ISBN-13: 978-94-011-7094-9
DOI: 10.1007/978-94-011-7094-9

Published in the USA by
MTP Press
A division of Kluwer Academic Publishers
101 Philip Drive
Norwell, MA 02061, USA

List of the submitted publications

I Heding LG. Determination of total serum insulin (IRI) in insulin-treated diabetic patients. Diabetologia 1972;8:260–6.

II Heding LG, Larsen UD, Markussen J, Jørgensen KH, Hallund O. Radioimmunoassays for human, pork and ox C-peptides and related substances. Horm Metab Res (Suppl) 1974;5:40–4.

III Heding LG. Radioimmunological determination of human C-peptide in serum. Diabetologia 1975;11:541–8.

IV Heding LG. Specific and direct radioimmunoassay for human proinsulin in serum. Diabetologia 1977;13:467–74.

V Heding LG. Insulin, C-peptide, and proinsulin in nondiabetics and insulin-treated diabetics. Diabetes 1978;27 (Suppl 1):178–83.

Preface and acknowledgements

The RIAs reported in the submitted 5 articles, as well as in other studies discussed in this overview, were developed at the Novo Research Institute. The work started in 1964 at a time when the state of art called for development of reliable routine methods for such assays. The insulin RIA was followed by C-peptide and proinsulin RIAs after Don Steiner's important discovery of proinsulin in 1967.

I am grateful to Dr Jørgen Schlichtkrull and the late Dr Knud Hallas-Møller for the excellent working conditions over many years. I appreciate their positive attitude to my interest in developing methods for basic diabetes research, such as for example the human proinsulin RIA, as well as methods applied for improving the purification of insulin to achieve optimal diabetes therapy.

The work could not have been performed without the cooperation of many colleagues and I am especially grateful to Jens Brange, Ole Hallund, Klavs Jørgensen and Jan Markussen, as well as to Dr Ole Faber who worked at the Hvidøre Hospital. Special thanks are due to my four excellent and ever-patient technicians, Majken P. Pedersen, Bente Hansen, Marianne Heiden and Lisbet Pedersen.

Also very important was the extensive international cooperation established during these exciting years with many excellent scientists, and Kumar Naithani, Arthur Rubenstein and Don Steiner deserve special mention in this respect.

It has been a pleasure and incentive to see the methods thus developed used internationally in diabetes research in collaboration with inspiring and enthusiastic researchers, among whom I should like to mention Robert Turner, Johnny Ludvigsson, Hans Åkerblom, Bengt Persson, George Alberti and David Owens.

One person deserves special mention: Jørgen Schlichtkrull, who showed great interest in my work and never ceased to give relevant and constructive criticism. I am grateful for the many long, committed, intense and invigorating dicussions we have had throughout the years.

Finally, sincere thanks to Professor Ulrik V. Lassen, who has given me strong encouragement to attempt presenting the essence of my work, to Kathleen Larsen whose consistent, invaluable linguistic input is gratefully appreciated, and to Merete Sørensen who undertook the typing with artistic skill and patience with the many corrections made.

Contents

Introduction

THE DISCOVERY OF INSULIN, C-PEPTIDE AND PROINSULIN, SPECIES DIFFERENCES

Crude insulin was extracted and isolated from dog pancreases removed 7–10 weeks after ligation of the pancreatic duct for the first time in 1921 by Banting and Best. Its lifesaving properties were subsequently documented in pancreatectomized dogs in 1922. Only about six months elapsed from its discovery until systematic collection of calf and ox pancreases and, later, porcine pancreases was established and the blood glucose lowering effect in diabetics using the crude extract from these glands proven. It was not until 1960, however, that the primary structure of all three species of insulin, human, porcine and bovine, was elucidated[107].

It then became evident that the differences between bovine, porcine and human insulin amounted to at most 3 amino acids. These apparently small differences have a significant impact on the physico-chemical characteristics of the three insulins which again affect, e.g., the rate of absorption of insulin preparations from the subcutaneous tissue[109] and their immunogenicity, i.e. the ability to induce insulin antibodies in diabetics as a side-effect of subcutaneous administration[66,120–122]. Furthermore, due to the species differences, the three insulins react differently with some insulin antibodies [138,140], and this has to be taken into consideration when such antibodies are used in an insulin radioimmunoassay (RIA). On the other hand, the great similarity between the three species has not permitted the establishment of a species-specific insulin RIA, as it has not been possible to produce useful antibodies reacting exclusively with each individual insulin.

Notwithstanding the introduction of purification by crystallization, new analytical methods such as disc electrophoresis proved that the crystals contained several impurities[101], which were estimated to constitute about 10–15% of once-crystallized insulin[123]. Gel filtration was used by Steiner and co-workers in 1968[129] to fractionate once-crystallized insulin into three peaks, the so-called a-, b- and c-components, named in descending molecular size. Further fractionation of the three peaks using other chromatographic procedures revealed that each peak contained a sizeable number of substances.

The year before, Steiner and his co-workers[126,127] had discovered the presence of the biosynthetic precursor of insulin in extracts of human β-cell tumours and isolated rat islets. The precursor had a higher molecular weight than insulin, reacted strongly with insulin antibodies and could be converted by trypsin to a substance eluting at the insulin position after gel filtration. The precursor was termed proinsulin. In once-crystallized bovine and porcine insulin one of the major impurities was found to be proinsulin[128] and the further characterization of this and other components in crystalline insulin was performed using bovine material[129]. It appeared that bovine proinsulin is a single chain peptide with a molecular weight of approx. 9000. Proinsulin contains the complete insulin molecule and a connecting peptide which links the N-terminal amino acid in the A-chain to the C-terminal of the B-chain. Proinsulin is converted into insulin, C-peptide and two pairs of basic amino acids by enzymatic cleavage in the β-cells. C-peptides are always 4 amino acids shorter than the corresponding connecting peptides, namely lacking the Arg–Arg and Lys–Arg sequences at the N- and C-terminal, respectively.

Proinsulins from different species differ not only in insulin moiety but also, to a much larger extent, in the C-peptide part of the molecule, where the differences between the three species are of the magnitude of 50%. In addition, the number of amino acids in C-peptide also varies, e.g. human C-peptide consists of 31 amino acids, porcine of 29, bovine of 26 and dog C-peptide of only 23[130]. The amino acid composition of bovine C-peptide was first determined as a part of the proinsulin molecule[129] and later the sequence was determined on C-peptides isolated in the salting out liquor from the insulin manufacturing process[96,133]. Due to this great variability antibodies raised against one species of C-peptide (or the C-peptide part of proinsulin) did not react with other species of C-peptide. Thus, in contrast to establishment of RIAs for insulins, each species of C-peptide and proinsulin requires its unique set of reagents.

The biological activity of proinsulin compared to that of insulin was found to be very low originally[129] when tested in an isolated fat cell system ($\sim 2\%$). In in vivo tests in the intact animal – mouse and rabbit – proinsulin exhibited a prolonged hypoglycaemic effect compared to insulin, rendering routine measurement of the bioactivity against the insulin standard invalid. Judged by the area under the blood glucose curves from 0 to 2–3 h after the injection, proinsulin retains 15–25% of the activity of an equimolar amount of insulin[120].

Although secreted into the blood, C-peptide has not been shown to possess any biological activity whatsoever. Its sole function appears to be ensuring the proper folding of the proinsulin chain[130].

The establishment of a radioimmunoassay

THE FIRST SYSTEMATICAL APPROACH

Yalow and Berson were the first to develop a RIA[139]. Naturally it happened to be for insulin, since the same researchers in 1956 had proven that conventional insulin therapy induced circulating insulin-binding antibodies in all patients treated, as demonstrated by the ability of their serum (and immunoglobulin fractions) to bind ^{131}I-labelled insulin[7]. A detailed report[139] outlined the essential requirements needed to establish this RIA: raising antibodies with sufficiently high affinity to permit a low detection limit of the assay and sufficiently high capacity for practical routine work, preparing a labelled substance with high reactivity towards the antibodies and, last but not least, to have available standards of the substance to be measured. For the first time guinea pigs were used to raise insulin antibodies and they have remained the animal of choice for the preparation of insulin antibodies in spite of their drawbacks – small blood volumes only obtainable by heart puncture or eye blood sampling. The outstandingly positive characteristics of their antibodies are the combined high affinity and capacity, allowing low detection limit, high sensitivity and high dilution of the antiserum. The latter characteristics may be a consequence of the great differences between the bovine, porcine and human insulin compared to guinea pig insulin (approx. 19 amino acids out of 51), rendering the exogenous insulins highly immunogenic in the guinea pig. Berson and Yalow's pioneer publications on RIA techniques in the years 1956–1960 were quickly followed by other researchers' variations with many improvements and this still continues. Two deserve mention here, namely the introduction of a preliminary extraction of serum to remove the endogenous insulin antibodies, which would interfere with the RIA, followed by a separation of free and bound ^{131}I-insulin using salt precipitation[41], and the use of a second antibody to separate the free and bound ^{131}I-insulin[44,102].

REQUIREMENTS FOR ANTIBODIES USED IN RIA

Raising antibodies with high affinity and capacity turned out to be problematic, especially for peptides with low molecular weight (e.g. glucagon and

C-peptide). However, the introduction of various ways of enhancing the immunogenicity of a molecule (adjuvants, coupling to larger molecules, etc.) has virtually overcome that problem[31,49,64,115]. The affinity and capacity are not the only characteristics which determine whether an antibody is suitable for a RIA. The specificity, i.e. the ability of an antibody to react exclusively with the substances to be measured, is a decisive factor influencing the selection of antibodies. In order to fully characterize a series of antibodies a number of fragments, derivatives and related substances are extremely helpful, e.g. as used for C-peptide and gastrin antibodies[31,115]. In conclusion, the selection of antibodies – whether polyclonal or monoclonal – remains a pragmatic compromise in which the following characteristics are taken into account: affinity, capacity, specificity and availability.

CHARACTERIZATION OF THE STANDARD SUBSTANCE AND THE SUBSTANCES TO BE MEASURED

In the early years of the RIA the standard substances used to produce the standard curve would by today's chemists be termed rather ill-defined. This mainly reflects the different stages of purification and analytical methods then compared to now. Advanced extraction and separation methods (preparative HPLC) and analytical characterization have been developed since the sixties and, in addition, methods like amino acid sequencing of a peptide can now be performed on microgram amounts. In their RIA, Yalow and Berson[139] used two different human insulin standards reported to have biological activities of 1.8 U/mg and approx. 6 U/mg, respectively. This means a purity of 5% and 20% compared to the pure human insulin available today. The composition of the remaining part of the two standards remains unknown. The standardization was performed using a bioassay which tends to underestimate the number of immunoreactive insulin moities, e.g. proinsulin and its intermediates. These substances react well in most insulin RIAs; consequently, the RIA standard value might have been considerably higher than the bioassay value. Underestimation of the H_2O content of a lyophilized standard which may exceed 50% of the protein content by weight represents another source of error, as does the presence of buffer substances such as $(NH_4)HCO_3$ in a lyophilized standard. It is remarkable how often an unexpected result can be explained by simple errors in weighing, N-analysis, amino acid standardization, dilution, etc. Preparation of purified microgram quantities of a substance from a source where its content amounts to less than 10^{-9} by weight, e.g. human proinsulin[76] poses special problems in developing refined methods of purification, analysis and characterization.

Standards prepared by peptide synthesis or semisynthesis are becoming increasingly available[103,141] and, gradually, biosynthetic peptides will become in common use in RIAs[15,35,98]. A comparison of such standards with their natural counterparts is obligatory prior to their general use in RIAs. Even after conventional analytical methods had failed to show a difference between a natural and synthetic C-peptide preparation, antibodies were able to pick up an error, e.g. in a synthetic C-peptide[54,103].

The stability of a standard varies greatly with substance and storage conditions – humidity, concentration, temperature, etc. Most proteins tolerate years of storage as a powder at $-20°C$ provided moisture is excluded. Many are likewise stable in solution when stored at $-20°C$. Stability in the presence of proteolytic enzymes is quite unpredictable and the use of reagents devoid of such enzymes strongly recommended.

The ideal situation – having a pure, well-defined standard used for measuring the very same substance in serum – is virtually non-existent. Standards are becoming increasingly well-defined and, at the same time, it is realized that the substances measured in biological fluids are normally a mixture of related substances, often the hormone together with a prohormone, fragments of the hormone, etc. Gel filtration may be used to separate the various substances, whereupon they can be measured and examined for dilution effect in the RIA[80]. Such methods are tedious and not suitable for routine analysis, and practical compromises are the alternative. In case of heterogeneity and dilution effects, the latter can normally be minimized by antiserum selection and by working in a certain range of the standard curve. For this very reason, the term *immunoreactivity* should be added whenever a RIA is used to measure a substance, e.g. *proinsulin-like immunoreactivity* (PLI) and *immunoreactive insulin* (IRI).

In conclusion, standards of peptides or proteins should be characterized by their purity, e.g. on HPLC (at least two different systems), amino acid composition and N- and/or C-terminal amino acid(s). Performed on the stock solution, these analyses give information of the molar concentration and then guide subsequent dilutions. If the standard substance is biologically active a bioassay is normally performed as well. However, when a substance like insulin is more than 99% pure and its bioactivity once determined, the chemical analyses are far more important due to their far greater accuracy[112]. Stability of the standard, both in dry and in dissolved, diluted state, should be monitored and serial dilutions of the biological fluids or extracts to be assayed carried out routinely to estimate any dilution effect in order to detect and minimize this potential problem.

THE TRACER, CHARACTERIZATION AND LIMITATIONS

The standard should be identical or closely related to the substance(s) to be measured, but this is not necessary for the immunogen that is used to induce the antibodies, nor for the substance to be radio-labelled and used as the tracer. The requirement to the tracer is that it reacts with the same antibodies as do the standard and the assayed substances, meaning that the standard and the tracer compete for the antibody sites.

Berson and co-workers[7] used ^{131}I with a $T_{\frac{1}{2}}$ of 8 days but, later, ^{125}I with a $T_{\frac{1}{2}}$ of 60 days became available and quickly became the radio-tag of choice. The oxidizing conditions under which the ^{125}I is chemically bound to one or several tyrosine residues in the insulin is of importance for the distribution of radioiodine within the insulin molecule. The iodate method[69,70] yields preferentially ^{125}I incorporation in tyrosine A19, also achievable with horseradish peroxidase[83], while using lactoperoxidase[70] results mainly in iodination at the A14-tyrosine position. Fractionation and purification of the iodination mixture are still obligatory and have an impact on the quality and stability of the tracer[70].

During purification it is essential not only to remove free radioiodide, which could interfere non-specifically later in the RIA, but also to remove molecules having more than one ^{125}I per molecule, because the decay of one ^{125}I destroys the molecule and leaves either ^{125}I or a fragment still ^{125}I-labelled that probably cannot be bound to antibodies but gives elevated non-specific background values. Mono-^{125}I-insulin (position 14 or 19 in the A-chain) has been shown to be of superior quality as a tracer for the insulin RIA compared to more crude ^{125}I-insulin preparations. The concentration of immunoreactivity declines as the ^{125}I decays due to immunochemical destruction of the parent insulin molecule, but the high quality and extremely low non-specific binding to glassware and proteins remains virtually unchanged[70]. Some peptides like the C-peptides do not have a tyrosine residue, hence tyrosine has to be added either by semisynthesis[51,95] or during the full synthesis procedure[104]. In conclusion, purification of the tracer from the iodination mixture must be ranked very high and, unless proven advantageous, peptides containing more than one ^{125}I per molecule should be removed to ensure optimal quality and storage properties of the resulting mono-^{125}I-peptides.

QUALITY AND REPRODUCIBILITY OF RIA

Setting up a RIA requires a series of qualified choices to be made after the availability and quality of three main reagents have been ensured, i.e. buffer system, preservative, pH, separation method, incubation time(s) and whether a one-step ('equilibrium', where standard or sample, tracer and antiserum added last are incubated mixed) or a two-step assay ('dis-equilibrium' or titration, where standard or sample are incubated with the antiserum first and the tracer added later to bind residual vacant antibody sites) should be preferred. In addition, it has to be decided whether proteolytic degradation may pose a problem during the assay or storage period, as is the case in the glucagon and C-peptide assays[32,36,48] and therefore require addition of suitable amounts of a protease inhibitor such as aprotinin to samples and/or standards.

A number of different separation methods, each having their advantages and drawbacks and individual sources of error, have been published; they date right back to the original chromato-electrophoresis of Yalow and Berson[139], later to become a quick, routine method[143], the double antibody method[44,102], the use of ion-exchange resin[99], salt precipitation[41], solid phase separation using insulin antibodies coated on the wall of plastic tubes[14], ethanol[45], activated charcoal[2], *polyethylene glycol* (PEG)[25], and numerous variants of these. The *enzyme-linked immunosorbent assay*, ELISA, is not referred to here, but will be mentioned separately in a later chapter. None of the mentioned methods can be recommended universally, but one common rule should be obligatory for all laboratories performing RIAs routinely, namely the daily quality control incorporated in each assay and recorded and checked centrally. This is stressed due to my experience in dealing with RIA problems from various laboratories where I have been surprised to learn how few there are who do take the limited extra time to carry out the necessary, continual checks of their results to ensure they are as uniform and reliable as possible.

Controlling tracer concentration before and after a RIA, adding a few standards at the end of the assay and analysing at least two reference sera (or other relevant samples) are minimum requirements. From the results of these controls it is possible to check drifting of the standard curve, reproducibility of the assay over the years, long-term stability of the standard solutions, etc.[62]. The nonspecific precipitation of the ^{125}I-labelled tracer by serum, plasma or extracts should be compared to that in the buffer. Serial dilutions of serum, binding of the tracer by a surplus of antibodies, etc. are other daily routine checks.

Thus, in conclusion, RIA results should be substantiated by including day-to-day and long-term control checks and such figures used objectively to determine when an assay should be rejected and repeated[62].

Background for development of improved and new RIAs for insulin, C-peptide and proinsulin

INSULIN

In the mid-sixties when it was gradually accepted that insulin could be measured fairly accurately in serum many laboratories established RIAs based on the principles of the pioneers. It became mandatory to improve several of the steps in order to obtain routine methods – especially the separation of free and antibody-bound insulin – to replace Berson and Yalow's chromato-electrophoresis. Secondly, there was a need for a routine method to extract and isolate insulin from serum of insulin treated diabetics having insulin antibodies (Ab_I). Most of the IRI in serum from such patients is bound to Ab_I while a minor part is free IRI. The extraction procedure should lead to isolation of the free and bound IRI clearly separated from the Ab_I. The bound insulin is closely related to the antibody level[61,121] and gives more exact information about the immunogenicity of the insulin preparations than the mere percentage of bound ^{125}I-insulin obtained after addition of a randomly selected amount of tracer[19,125]. In addition, it became of interest to characterize the total IRI circulating which could, in theory, consist of a mixture of insulins and proinsulins[56].

C-PEPTIDES

After the recognition that C-peptide was not only being stored equimolarly but, in addition, secreted in equimolar amounts to insulin[51], the interest in establishing a sensitive method for human C-peptide determination grew as the potential use of C-peptide as a marker for endogenous insulin could give new and exciting information about the endogenous insulin secretion in insulin-treated diabetics. Thus, C-peptide values represent an objective evaluation of the insulin secretion and the method of choice for studying the effect of alternative therapies on preserving, restoring or even curing the β-cell defect in insulin-treated diabetics. An assay developed for porcine C-peptide is useful whenever the pig is to be used in studies replacing or preceding human investigations.

PROINSULINS

Rubenstein and co-workers[100,118] early described the presence of a substance in serum with a higher molecular weight than insulin but, at the same time, containing an insulin moiety as judged by RIA. The need for a RIA for each species of proinsulin was due to interest in determining proinsulin in serum from diabetics and non-diabetics and the necessity to have sensitive RIAs to monitor the purification and subsequent removal of porcine and bovine proinsulin from insulin once it was discovered that the proinsulin in commercial insulin preparations was highly immunogenic[120]. It was intended to reduce proinsulin to such levels that no biological side effect was due to its presence[123]. The discovery of proinsulin gave rise to hopes of it giving clues to the pathogenesis of diabetes, e.g. incomplete cleavage in the β-cells, abnormally high secretion of non-converted proinsulin, etc. and this has been studied to some extent during the last 10 years.

Discussion of the methods

INSULIN

Characteristics, principles and sources of error

The insulin RIA was in very great demand right from its establishment and has remained so. The production of antibodies appeared to be very simple, no coupling of the antigen was required, and the first choice of experimental animal, the guinea pig, was fortunate. Minor variations in the immunization procedure regarding dose, frequency, adjuvant, species, purity and formulation of the antigen have been described, but are of slight significance, since it is possible to achieve acceptable antibodies of high capacity and high affinity by immunizing guinea pigs monthly with insulin[50,143]. Besides the capacity of the antibodies, very little characterization is normally considered important, except calculating the detection limit (the lowest amount of insulin which can be distinguished from the zero point, $p < 0.01$) and the reproducibility (the day-to-day variation).

Several standards have been in use in the insulin RIA, stemming back to 4–5–times crystallized human, porcine and bovine insulin, later replaced by monocomponent (MC) quality around 1970[120]. The mean biological potency of all three species of MC insulin is today 184 IU/mg N or 28.4 IU/mg of dry insulin ($\sim 5\%$ of water), using the mouse convulsion assay[112]. Today, the human pancreas derived human insulin has been replaced by human insulin made by transpeptidation of porcine insulin[97] or produced by fermentation of genetically manipulated microorganisms[15,98].

The very same substances as used for standards have been used for iodination with ^{125}I exclusively[69,70]. The quality of the tracer has been steadily improved to reach today's level of mono-A19 or A14 ^{125}I-insulin with a great advantage for the RIA using ethanol precipitation, namely low and constant non-specific precipitation of the non-antibody bound ^{125}I-insulin.

It is important to decide whether the RIA should be an equilibrium one-step assay or a dis-equilibrium two-step assay. Both methods have advantages and disadvantages. With the one-step assay a shorter total incubation time is required as the standard or sample is mixed with the tracer and then incubated for one period (4–24 h) with the added Ab_1. Using this procedure it is often possible to distinguish between closely related antigens, e.g. insulin from different species will compete differently with the tracer for the

Ab_I-sites. However, it is not always desired to aim for this characteristic, hence the two-step assay offers an alternative. Guinea pig insulin antibodies have a high binding affinity for insulin and it has been reported to take weeks to obtain equilibrium between preformed insulin-Ab_I complexes and added insulin[75]. A consequence of this is that once a complex between insulin and Ab_I is formed, the Ab_I now ligated to insulin is no longer available for reaction with other insulin molecules within the incubation periods. The two-step assay can therefore be used as a 'titration assay', meaning that the number of Ab_I-sites that were not occupied by binding of standard insulin or serum IRI in the first step can be quantified by their binding capacity to ^{125}I-insulin in the second step without an intermediate separation procedure. This type of assay is suitable to measure a sum of insulin immunoreactive substances with different affinities to the antibodies and is therefore preferred to the one-step assay. Examples are mixtures of human, porcine and bovine insulin, e.g. in total IRI from some insulin-treated diabetics.

The desired properties of the separation procedure are to have 100% of the free tracer in one phase and the antibody-bound tracer to be found 100% in the other phase. Even from a theoretical point of view this is not feasible, as the separation itself may induce disturbance of the equilibrium. Ethanol proved capable of precipitating all antibody-bound insulin compared with the reference methods, the double-antibody and paper chromatographic method[50]. Likewise, the fraction of free ^{125}I-insulin remaining in the supernatant was the same with the three methods when use was made of 96% ethanol yielding a final concentration of 79%. As for other separation methods, the protein concentration in serum and standards should be similar to get the same co-precipitation of free ^{125}I-insulin. The co-precipitation is also influenced by the quality of the tracer, temperature of the final mixture and ethanol quality. Degradation of ^{125}I-insulin by proteolytic enzymes present in the commercial albumin has by and large been eliminated by using purified albumin, e.g. produced by Behringwerke[50]. A specific requirement for ethanol separation – and a frequent source of error – is the presence of a minimum of 0.02 mol/l monovalent anions, such as Cl^-, in the incubation mixture[46]. Failure to ensure this results in the total absence of a precipitate upon ethanol addition.

Among the advantages of using ethanol for the separation procedure can be listed that it is a quick, easy and cheap process, even compared to the later, advantageous, PEG precipitation (greasy, high viscosity fluid)[25], and that the 96–97% ethanol can be purchased according to specifications. It is also in logical concordance with the total IRI extraction method where ethanol so far is the reagent of choice for obtaining reproducible results. The absence of the special problems related to the double-antibody precipitation is also an advantage. On the other hand, the method requires a tracer of high quality in order to minimize non-specific co-precipitation and cannot be used universally for all RIAs, in contrast to the double-antibody method. The dependence on monovalent anions and a standardized protein concentration may be regarded as a slight inconvenience.

In the sixties only Grodsky[41,42] had attempted to extract and analyse insulin from serum of diabetics and animals having Ab_I and only in a limited number

17

of samples, due to the complexity of the procedure. Later Ohneda and coworkers[108] used another complicated extraction procedure aimed at separating the total IRI from the Ab_1. The separation was, however, not complete. A more logical approach was reported by Pearson and Martin in 1970[110] who performed gel chromatography after adjustment of the serum to pH 3. Obviously, this approach is not suitable for routine use.

The total IRI method is based on the principle that antigen–antibody complexes formed at pH 7.4 dissociate immediately at pH $\sim$2.5 and do not recombine at pH 7.4, provided the solution medium is inhibitive for such reactions as shown for 75% ethanol (vol/vol)[50]. Later the use of PEG for isolation of free and total IRI was described, as PEG appeared to be highly selective in precipitating mainly immunoglobulins and complexes thereof while leaving most other proteins in the supernatant[79,105]. This principle was initially applied successfully to the RIA area, where 25% PEG was used as the precipitating agent[25]. The PEG method has the advantage over the ethanol total IRI method that no evaporation, such as that of the 75% of ethanol extract, is necessary. The final 12.5% PEG–serum mixture containing the total IRI is used direct in the RIA, the first step being a reaction with insulin antibodies at a PEG concentration of 6.25%. It then seemed questionable whether the insulin and Ab_1, which are dissociated at pH $\sim$2.5 prior to the addition of PEG and subsequently neutralized with NaOH, would not recombine to some extent in the presence of 12.5% PEG. Total IRI was determined in sera from 5 diabetic patients with Ab_1 using the ethanol and the PEG method, respectively. The total IRI resulting appears in Table 1. The 'total IRI' estimates obtained using the PEG method were clearly lower than the results from the ethanol method. In another experiment the recovery of ^{125}I-insulin added to a serum containing Ab_1 was subjected to a total IRI extraction using PEG. When the neutralized 12.5% PEG-serum mixture was allowed to stand 15 minutes prior to centrifugation, the recovery of the ^{125}I-insulin in the supernatant dropped from 94.3% to 81.5%. These results support the hypothesis that insulin and Ab_1 do react at pH 7.4 in the presence of 12.5% PEG and, consequently, ethanol must be preferred for the determination of total IRI.

Table 1 Total IRI (μU/ml) in diabetic sera using ethanol and PEG

Serum no.	Ethanol	PEG
3	221	171
38	992	700
51	6	6
67	1024	812
73	48	40

The presence of PEG in the RIA incubation mixture was found not to interfere in the assay[105], but others have found slight[79] or even strong interference[59,136] and recommended that the standard solutions should have the same PEG concentration as the samples. The 12.5% final PEG concentration recommended to achieve 100% precipitation of immunoglobulins is sufficient in most cases. Lately, however, we have observed that samples containing high amounts of Ab_1 often have residual Ab_1 in the PEG supernatant after a neutral PEG precipitation for the determination of free IRI. This could be eliminated by increasing the PEG concentration to 15% and adjusting the pH of the serum samples to 7.4[136].

The validity of free IRI determinations, including the most correct way of achieving such an estimate, is a lengthy discussion that will not be made here. It suffices to say that whether immediate bedside sampling in PEG[21] or the usual routine collection of blood processed to serum, which is stored at $-20°C$[37] followed by thawing and 2 h incubation at 37°C, is the most realistic remains controversial. On the other hand, the PEG procedure yields close to the theoretically correct value of free ^{125}I-insulin when compared to the new gel method developed by Jørgensen[71] and free IRI values may lead to valuable information[16].

The reactivity of proinsulin in the insulin RIA varies (70–100%) depending on antiserum and assay conditions. This leads to slightly overestimated values of serum insulin, due to the co-determination of proinsulin, hence the term IRI.

Usefulness

The insulin RIA and the total IRI method do not have any diagnostic importance, but have remained mainly of scientific value. The understanding of insulin secretion in type II diabetics, obesity, pretreatment investigations of type I diabetics, and many other diseased vs. normal state conditions has been enlarged and this may eventually lead to improvements in therapeutic approaches. The insulin RIA, besides its clinical use, is a daily tool in insulin manufacture and optimization of purification processes. The total IRI, which has to be performed whenever Ab_1 are present, continues to give information about the immunogenicity of exogenously administered insulin. The author is of the opinion that it gives the most exact measure of the activity of Ab_1 present and it is the only Ab_1-related method giving quantitative results. The implication of having elevated total IRI for the risks of hyperglycaemia and delayed hypoglycaemia was originally outlined by Berson and Yalow[8] and is currently under investigation as the preliminary reports[17,28,93] indicated a strong relationship.

The neutral ethanol precipitation[47] did not prove useful for the determination of free insulin in the presence of low affinity Ab_1[4]. Finally, it should be mentioned that the ethanol precipitation of insulin-antibody complexes after immunization with insulin has become an easy control method to register and rank the immunogenicity of various insulin preparations[121].

C-PEPTIDES

Characteristics, principles and sources of error

The development of the C-peptide RIAs was faced by considerable problems: first the preparation and purification of the standard substances, secondly, the antibody production and characterization and, thirdly, the introduction of tyrosine in the C-peptide.

Antibodies against all three C-peptides (Ab$_C$) were initially produced in guinea pigs by using the more readily available porcine and bovine proinsulin and crude human b-component obtained from first crystals of insulins resulting in a mixture of Ab$_I$ and Ab$_C$[51]. Others have used similar approaches[78,100] although connecting peptide, C-peptide coupled to albumin, and rabbits and goats have also been used[6,10,72,77,100]. The specificity for the C-peptide moiety was obtained by using ^{125}I-tyrosyl-C-peptide as the tracer rather than ^{125}I-proinsulin[51]. Removal of Ab$_I$ by a solid phase insulin prevented insulin and proinsulin from reacting in the C-peptide RIA, when ^{125}I-proinsulin was the tracer[1,142]. The most successful production of antibodies was obtained using synthetic human benzyloxycarbonyl-C-peptide coupled to albumin[31,103].

The shortage of natural human C-peptide and, consequently, the tracer and antibodies prompted the full synthesis of two batches of C-peptide, the benzyloxycarbonyl derivative and the tyrosyl-C-peptide for iodination[103,104]. The replacement of the natural pancreatic C-peptide by a synthetic one revealed a problem, as the two synthetic batches prepared reacted differently with some of the antibodies and selection of batch 2 was based on the observation that this batch gave the same dilution curves in the RIAs as a pancreatic extract containing high amounts of natural C-peptide[54,103]. Unfortunately, this synthetic C-peptide was not available in sufficient quantities to make it the reference material available from the National Institute for Biological Standards and Control, Hampstead, London, U.K. and, instead, the less attractive Lys(FOR)^{64}, 31–65 human proinsulin fragment was chosen[13]. However, recently a correct synthetic C-peptide has become commercially available from, e.g. Bachem Feinchemikalien AG, Switzerland, and a new human C-peptide made by genetic engineering is currently being evaluated in a number of laboratories (Bente Tronier, personal communication).

Since both C-peptide and proinsulin are present in serum and other biological fluids the question arises: to what extent does the presence of proinsulin interfere with the C-peptide determinations? Gel filtration of serum has been used to separate C-peptide and proinsulin and the fractions were assayed in a C-peptide RIA. It was found that proinsulin was present in much lower amounts than C-peptide and reacted less well in the C-peptide RIA than C-peptide[119]. A thorough systematic evaluation of the reactivity of proinsulin in seven C-peptide RIAs was performed and confirmed that the error in C-peptide values caused by co-determination of proinsulin could be regarded as negligible in most sera[32]. Correct C-peptide estimates have to include a separation from proinsulin and, as the combined gel filtration and RIA methods were too cumbersome, binding the proinsulin to guinea pig Ab$_I$, thereby preventing it from reacting in the C-peptide RIA, proved an attractive alternative. Initially, the Ab$_I$ was added as anti-insulin guinea pig serum[51] but, more logically, the binding to Sepharose-coupled Ab$_I$ (S-AIS) was developed[52], whereby the S-AIS proinsulin could be removed by centrifugation, resuspended and assayed in the solid phase[55] and C-peptide assayed in the solution without interference from proinsulin.

The separation of free and bound ^{125}I-Tyr-C-peptides using ethanol as for insulin RIA proved simple and the labelled C-peptides show very low non-specific co-precipitation in the presence of serum[52]. The sources of error are, besides the ones in common with the insulin RIA, mainly the presence of C-peptide fragments, degradation of C-peptide if stored improperly, presence of proinsulin and of non-specific immunoreactive substances in serum. The latter problem may lead to spuriously elevated serum C-peptide values and an increased day-to-day variation[52,54]. Thus antisera have to be selected to give the lowest value of C-peptide in patients without β-cell function. A major part of the variation in C-peptide normal fasting values reported with different antisera over the years can most likely be explained by lack of selection in this criterion[32].

The heterogeneity of serum C-peptide may cause non-linear dilution curves depending on the RIA system[78,79]. This problem has not been solved. Degradation in plasma depends on temperature and duration of storage[32,36]. Some antibodies react well with the degradation products hence rendering the degradation less important. Nevertheless prevention of degradation by addition of aprotinin is highly recommendable[32] and it remains to be confirmed whether a storage temperature of $-70°C$ is needed in order to avoid degradation completely[36].

In insulin treated diabetics with Ab_I, these antibodies bind insulin as well as proinsulin, prolong their half time and cause accumulation reaching concentrations often exceeding those of C-peptide[54,56,57]. Thus, in sera from individuals with Ab_I it must be recommended to remove the Ab_I-bound proinsulin by, e.g. PEG precipitation, followed by determination of C-peptide in the supernatant[79,136].

Usefulness

Several review articles have appeared praising C-peptide as a blessing on the part of nature, enabling the study of the natural history of the β-cell destruction in insulin-requiring diabetics and therapeutic measures investigating prevention or arrest of this process[9,12,33,67,92]. The basic finding that insulin and C-peptide are present in equimolar amounts in pancreatic extracts and are both secreted into the portal vein system[51,68] was followed by reports of higher molar concentrations of C-peptide in peripheral blood in humans, pigs and oxen[10,51], although the Rubenstein group originally reported equimolar amounts in peripheral blood too[100,119]. The absent or minimal uptake of C-peptide by the liver[82,113], and its approx. 7-fold higher peripheral levels than insulin[53] make this peptide an attractive marker of insulin secretion. The method has remained in high demand in diabetes research and insulin secretion studies in a broader sense, but is only of limited value for diagnostic purposes[90]. Thus, C-peptide has been used to study the short-term benefit of intensive therapy at onset[91], to show the importance of minimal residual endogenous insulin secretion[89], the functioning of pancreatic transplants[43,144], to evaluate immunosuppressive therapy[30,86,106,131], plasmapheresis[88], the β-cell function in newborn babies of diabetic mothers with Ab_I[11,60,111] and, above all, to investigate the natural history of β-cell function and destruction, especially in diabetic children[63,84,85]. The porcine C-peptide RIA has gained increasing interest as the pig is used in insulin secretion studies[34,82].

PROINSULINS

Characteristics, principles and sources of error

Soon after its discovery proinsulin was found in human serum as a molecule with insulin immunoreactivity eluting from columns of Sephadex G50 at the position of bovine proinsulin[39,100,116,117]. Analyzing fractions of calf serum extracts using a specific immunoassay for bovine proinsulin, measurable immunoreactive material appeared at the position of proinsulin[142]. This RIA was, however, unable to determine C-peptide which should have been present in far higher amounts, therefore the validity of the observation is dubious. The time- and sample-consuming, besides inaccurate, combined gel filtration and RIA methods prompted development of a quick and apparently attractive method using an enzyme claimed to degrade insulin only[27]. However, it appeared that the degradation of insulin was not complete and, in addition, that proinsulin was also degraded to some extent[22]. As a natural consequence of performing the separation of C-peptide from proinsulin bound to S-AIS prior to analysing for C-peptide, the testing of the immunoreactivity of the solid phase bound proinsulin in the C-peptide RIA[55] was performed. It appeared that the bound proinsulin had high molar reactivity compared to C-peptide[54,55]. Proinsulin is thus characterized as a molecule having both an insulin and a C-peptide immunoreactive moiety and neither insulin nor C-peptide reacts in the RIA. Apart from being more time-consuming, the proinsulin RIA is fairly robust without significant loss in proinsulin immuno-reactivity in stored samples, about 100% recovery, no dilution effects, etc. A special requirement to the proinsulin RIA is a lower detection limit as compared with, e.g. the C-peptide RIA, due to the much lower fasting concentration present in non-diabetics. Using an antibody against human benzyloxycarbonyl-C-peptide coupled to albumin[31] has further improved the detection limit, at the same time yielding virtually identical serum values as the previous antibody[58].

The first human proinsulin standard was prepared by or in cooperation with Rubenstein's group in Chicago[81,118] from more than 1 g of once-crystallized human insulin, resulting in less than 1 mg of standard. This task was extremely difficult, as HPLC for purification and characterization was not available, and methods requiring more material could not be used due to the limited amount available. For about 10 years this standard was distributed worldwide and used in the few routine proinsulin RIAs and for determination of the reactivity of proinsulin in the various C-peptide RIAs. Two new batches of proinsulin from human pancreases were recently prepared and compared with the proinsulin made by recombinant DNA technology by Lilly[35] and the hitherto-used standard[76]. The new standards were characterized using HPLC, quantitative amino acid analysis and partial amino acid sequencing. The three new standards yielded identical standard curves in two totally different RIAs in three different laboratories and were found to be three times stronger than the original standard. This is hardly due to instability, since quality control of the old standard had been carried out through approx. 10 years[76]. In conclusion, this investigation clearly demonstrated that the old standard is obsolete and should be replaced by a new one. The reason for the observed difference is yet to be elucidated.

The pressing question is: to what extent does this change in standard influence previous results and conclusions? Proinsulin RIAs have not been widely used, mainly due to the shortage of a standard, hence results are scarce. A direct consequence of using a new standard is that all previous results obtained with the method of Heding[55] have to be divided by a factor of 3. Ranking of proinsulin content in sera and conclusions about the abnormal levels in diabetes and other diseased states compared to normals remain the same. Rainbow and co-workers[114], using the same standard but a different RIA with rabbit Ab_C having a similar detection limit, found virtually the same fasting mean values in non-diabetics and maturity onset diabetics, as reported by Heding[55], and this further substantiates that previously reported proinsulin results should be divided by 3. However, calculations of the molar ratios between proinsulin and insulin and proinsulin and C-peptide, respectively, must be considered incorrect[59,63,87] and subjected to re-evaluation.

The RIAs for porcine and bovine proinsulin[1,18,23,56,132,134] for the determination of proinsulin in crystallized insulin and extracts of serum from diabetic sera containing Ab_I do not include a separation of C-peptide, since there are no bovine and porcine C-peptide present.

Recent approaches to make the proinsulin RIA more readily applicable suffer from low sensitivity[20,24,40], necessitating 'up concentration' of proinsulin-like material from serum in order to obtain results above the detection limit of the RIA. However, it is encouraging that antibodies specific for proinsulin, i.e. not reacting with insulin or C-peptide, have been reported[20,24] and recent ELISA approaches are also promising[26].

The prevailing source of error – apart from those mentioned earlier, including the presence of Ab_I, which necessitates an acid extraction prior to analysis – remains the heterogeneity of proinsulin-like material in serum and its consequences for the RIA results. Besides proinsulin, once-crystallized insulin extracted from bovine and porcine pancreas contains nearly the same amount of intermediates cleaved either at the junction between the connecting peptide and the A-chain or the B-chain as well as a series of related products, which are most likely present in serum too. The recent observation by Gray and co-workers[40], that biosynthetic human proinsulin was found to be at least one hundred times less potent in their proinsulin RIA than extracted human pancreatic proinsulin, could be explained by lack of reactivity of intact proinsulin with the Ab_C employed and indicated that the extracted proinsulin was rather one or more of the intermediates. This has never been investigated in the human proinsulin RIA, while in the bovine and porcine proinsulin RIAs the intermediates reacted similarly to proinsulin and C-peptide, indicating that the Ab_C was directed towards the C-peptide part of the molecule[51]. Characterization of the proinsulin-like material secreted from human pancreases is needed to clarify this question using a combination of gel filtration and RIA[38] or well-defined monoclonal antibodies[94].

Usefulness

It is incomprehensible why the prohormone, proinsulin, is secreted from the β-cell in humans but not in the closely related species, the pig (own observation). It has been found that in the diabetic state proinsulin often constitutes a higher proportion of the IRI than in non-diabetics[29,55,87], however due to the excessive hyperglycaemia a qualified conclusion is difficult to draw, since normal rat β-cells have been reported to degranulate and secrete higher than normal levels of proinsulin under prolonged hypersecretion[135]. Other conditions, like hyperthyroidism[124] and cirrhosis[73], also seem to be accompanied by increased proinsulin secretion.

Thus serum proinsulin measurements in individuals without Ab_1 have scientific value and should be performed whenever doubt can be expressed about the composition of the IRI. In individuals with Ab_1 the human proinsulin values found after acid extraction no longer reflect the secretion of proinsulin, but are dominated by the altered metabolic clearance due to the prolonged half time of the Ab_1–proinsulin complexes. The presence of bovine and porcine proinsulin in addition to human proinsulin in patients treated with, e.g. five times recrystallized insulin of mixed species is a consequence of the impure insulin and its immunogenicity[56]. The RIAs for bovine and porcine proinsulin have been applied mainly for improving the purification of commercial insulin and were later introduced as a quality control by the FDA[18]. These methods were a prerequisite for the many innovations made in order to produce the highly purified monocomponent insulin.

Finally, measurement of fasting serum levels of proinsulin in insulinoma-suspected patients has proven of diagnostic value[58,65,137] and relieved many patients of the difficult and less certain stimulation or suppression test used hitherto. The validity of the fasting proinsulin has been evaluated in 206 insulinomas[65] and, although absolute values are to be divided by 3, the diagnostic value of a fasting proinsulin determination remains incontrovertible. In neonates with nesidioblastosis hyperproinsulinism also persists[5].

The present state and expected future development of methods for the determination of β-cell secretory products

The insulin and C-peptide RIAs possess reproducibility and precision and are suitable for routine use, whereas the human proinsulin RIA remains more complicated. New antibodies with improved qualities continue to appear and the rabbit anti-C-peptide serum now introduced in a new kit from Novo is an example of a development where the main purpose was to secure long-term continuity in supply combined with high specificity and reproducibility of this widely used reagent. RIAs have maintained their position in most laboratories in spite of the promising ELISAs originally described about 10 years ago. Their detection limit was not up to that of RIA[3,74] and the benefit of avoiding ^{125}I is not without cost, as some of the ELISA reagents are not entirely harmless. On the other hand, the long shelf-life of the enzyme reagent, the speed and capacity are a great advantage, and a most promising ELISA for proinsulin was recently described[26], in which the ever-present non-specific serum interference seems to have been solved.

The introduction of monoclonal antibody technology with the possibility of producing large quantities of specific, well-defined antibodies has led to some use in RIA and ELISA systems, especially for characterization of antigens[94,138], although affinity and stability of the antibody seem to remain a problem.

Thus, although capacity, low cost and high speed are desirable, introduction of new methods will require that they are at least as sensitive and reproducible as the RIAs. Consequently, the future will probably not see an 'either/or' use of RIA *vs*. ELISA or polyclonal *vs*. monoclonal antibodies, but a practical combination of all whenever the quality of an assay can be significantly improved.

Summary and conclusion

RIAs have been developed for determination of the three main secretory products of the β-cells: insulin, C-peptide and proinsulin using ethanol for the separation of free and antibody-bound peptides. Besides measuring immunoreactive insulin (IRI) direct in serum, a routine method for total IRI using acid ethanol to isolate all IRI from serum of insulin-treated diabetics with insulin antibodies was established. The total IRI should not be compared with IRI values from sera devoid of antibody, but should be regarded as a measure of the binding characteristics of the antibodies. The molar reactivity of proinsulins in the insulin RIA varies from approx. 70% to 100%, depending on antiserum and assay conditions, while C-peptides show no reactivity. This introduces a small overestimation of serum insulin (hence IRI) in humans.

C-peptide RIAs were developed for the three species and the human and porcine C-peptide RIAs have been applied widely for serum determinations. Presence of immunoreactive substances in serum unrelated to C-peptide necessitated selection of the C-peptide antibodies which gave the lowest values in serum from patients without C-peptide. Degradation was minimized by addition of aprotinin and subsequent storage at $-20°C$. Proinsulin reacts to varying degrees in the C-peptide RIAs (30–100%). The low molar ratio between proinsulin and C-peptide in most human sera without insulin antibodies limits the error of co-determining proinsulin on the C-peptide values. A routine method for removing the proinsulin was developed to be used whenever exact C-peptide determinations are to be performed. Diabetics with insulin antibodies may have proinsulin (also of animal origin) in amounts exceeding those of C-peptide. Therefore removal of the antibody–proinsulin complexes is mandatory prior to a C-peptide determination in order to avoid overestimation of the endogenous insulin secretion.

Determination of human serum proinsulin was performed as a two-site RIA, the first step being a binding of proinsulin to solid phase insulin antibodies, followed by determination of the C-peptide moiety of the solid phase bound proinsulin in a C-peptide RIA. Neither C-peptide nor insulin reacts in the human proinsulin RIA.

Besides being used as a research tool to measure the β-cell secretory products in serum in normal and diseased states, primarily diabetes, the RIAs for insulin and the bovine and porcine proinsulin have been useful in optimizing yields and purification processes in the manufacture of insulin.

26

Resume og konklusion

Formålet med det foreliggende arbejde har været at udvikle radioimmunoassays (RIA) til bestemmelse af de tre vigtige peptider, som secerneres
af β-cellerne: insulin, C-peptid og proinsulin. Anvendelse af ethanol til separation af frie og antistofbundne peptider blev indført som et enkelt alternativ
til de etablerede teknikker.

Udover målinger af immunoreaktivt insulin (IRI) direkte i serum, er
der desuden udviklet en rutinemetode til bestemmelse af total IRI med
anvendelse af sur ethanol til isolering af al IRI i serum fra insulinbehandlede
diabetiske patienter med insulinantistoffer. Værdier for total IRI bør ikke
direkte sammenlignes med IRI værdier fra sera uden antistoffer, men
betragtes som en målestok for antistoffernes evne til at binde insulin. Pro-
insulinernes molære reaktionsevne i insulin RIA varierer fra ca. 70% til
100%, afhængig af antiserum og metodik, hvorimod C-peptid ikke reagerer.
Dette medfører at bestemmelsen af serum insulin hos mennesker giver lidt
for høje værdier, og derfor anvendes betegnelsen IRI i stedet for insulin.

Der er udviklet C-peptid radioimmunoassays for de tre arter, okse, svin
og human; human og svine C-peptid radioimmunoassays har vundet vid
udbredelse til serumbestemmelser. Det blev påvist, at serum indeholdt immuno-
reaktive substanser uden relation til C-peptid, som reagerede i varierende
grad med forskellige antistoffer og dermed gav falsk høje værdier. Derfor
blev de C-peptidantistoffer, som gav de laveste værdier i serum fra patienter
uden C-peptid, udvalgt til RIA. Proinsulin reagerer i varierende grad i C-
peptid radioimmunoassays (30–100%). I de fleste serumprøver fra mennesker
uden insulinantistoffer er det molære forhold mellem proinsulin og C-peptid
så lavt, at den fejl, der opstår ved medbestemmelse af proinsulin i C-pep-
tidanalysen, er uden betydning. Der blev dog udviklet en rutinemetode til
fjernelse af proinsulin i de tilfælde, hvor nøjagtige C-peptid-bestemmelser skal
udføres. Diabetikere med insulinantistoffer kan have værdier for proinsulin
(human og animalsk bundet til antistof), der overstiger C-peptidværdierne.
Derfor er det påkrævet, at fjerne proinsulin antistofkomplekser forud for en
C-peptidbestemmelse med henblik på at undgå for høje bedømmelser af den
endogene insulinsekretion.

Bestemmelsen af humanproinsulin blev udført som et to-trins RIA; det
første trin bestod i at binde proinsulin til 'solid phase' insulinantistoffer,
efterfulgt af bestemmelse af C-peptiddelen af det 'solid phase'-bundne proinsulin i et C-peptid RIA. Hverken C-peptid eller insulin reagerer i dette
humanproinsulin RIA.

27

De omtalte RIA har hovedsageligt været anvendt som forskningsredskab til at bestemme insulin, C-peptid og proinsulin i raske og syge personer – først og fremmest diabetikere. Derudover har insulin og okse- og svineproinsulin metoderne fundet anvendelse ved optimering af udbytter og renhed af insulin ekstraheret fra pankreas, og humanproinsulin RIA til at verificere en insulinom diagnose.

References

1. Aaby P. Concentration of porcine proinsulin-like material in plasma of insulin-treated diabetics in relation to purity of insulin preparations. Horm Metab Res 1979;11:455–7.
2. Albano JDM, Ekins RP, Maritz G, Turner RC. A sensitive, precise radioimmunoassay of serum insulin relying on charcoal separation of bound and free hormone moieties. Acta Endocrinal (Copenh) 1972;70:487–509.
3. Angelo L, Patkar SA, Tronier B. A highly sensitive monoclonal antibody against human C-peptide. Diabetes Res Clin Pract 1985;1(Suppl 1):18–9.
4. Asplin CM, Goldie DJ, Hartog M. The measurement of serum free insulin by steady-state gel filtration. Clin Chim Acta 1977;75:393–9.
5. Aynsley-Green A, Jenkins P, Tronier B, Heding LG. Plasma proinsulin and C-peptide concentrations in children with hyperinsulinaemic hypoglycaemia. Acta Paediatr Scand 1984;73:359–63.
6. Beischer W, Keller L, Maas M, Schiefer E, Pfeiffer EF. Human C-peptide. Part 1: Radio-immunoassay. Klin Wochenschr 1976;54:709–15.
7. Berson SA, Yalow RS, Bauman A, Rothschild MA, Newerly K. Insulin-I[131] metabolism in human subjects: Demonstration of insulin binding globulin in the circulation of insulin treated subjects. J Clin Invest 1956;35:170–90.
8. Berson SA, Yalow RS. Studies with insulin-binding antibody. Diabetes 1957;6:402–7.
9. Binder C, Faber O. C-peptide and proinsulin. In: Alberti KGMM, Krall LP, eds. The Diabetes Annual/1. Amsterdam, New York, Oxford: Elsevier, 1985:406–17.
10. Block MB, Mako ME, Steiner DF, Rubenstein AH. Circulating C-peptide immuno-reactivity. Diabetes 1972;21:1013–26.
11. Block MB. Pildes RS, Mossabhoy NA, Steiner DF, Rubenstein AH. C-peptide immuno-reactivity (CPR): A new method of studying infants of insulin-treated mothers. Pediatrics 1974;53:923–8.
12. Bonser AM, Garcia-Webb P. C-peptide measurement and its clinical usefulness: A review. Ann Clin Biochem 1981;18:200–6.
13. Caygill CPJ, Gaines Das RE, Bangham DR. Use of a common standard for comparison of insulin C-peptide measurements by different laboratories. Diabetologia 1980;18:197–204.
14. Ceska M, Grossmüller F, Lundkvist U. Solid-phase radioimmunoassay of insulin. Acta Endocrinol (Copenh) 1970;64:111–25.
15. Chance RE, Kroeff EP, Hoffmann JA. Chemical, physical, and biological properties of recombinant human insulin. In: Gueriguian JL, ed. Insulins, Growth Hormone and Recombinant DNA Technology. New York: Raven Press, 1981:71–86.
16. Chantelau EA, Sonnenberg GE, Best F, Heding LG, Berger M. Target fasting glycaemia for pump-treated type-I diabetics. Klin Wochenschr 1984;62:328–30.
17. Chantelau E, Sonnenberg GE, Heding LG, Berger M. Impaired metabolic response to regular insulin in the presence of a high level of circulating insulin-binding immunoglobulin G. Diabetes Care 1984;7:403–4.
18. Chiu YH, Gueriguian JL. Radioimmunoassays for determination of proinsulin content in purified insulin crystals. In: Hormone Drugs. Proceedings of the FDA–USP Workshop on Drug and Reference Standards for Insulins, Somatropins, and Thyroid-axis Hormones.

Bethesda, Maryland, May 19–21, 1982. Rockville, Maryland: United States Pharmacopeial Convention, 1982:216–25.

19. Christiansen AH, Heding LG. Quantitative radioimmunoelectrophoresis for determination of binding to IgG of insulin and other polypeptides. Scand J Immunol 1983;17(Suppl 10):273–8.
20. Cohen RM, Nakabayashi T, Blix PM, et al. A radioimmunoassay for circulating human proinsulin. Diabetes 1985;34:84–91.
21. Collins ACG, Pickup JC. Sample preparation and radioimmunoassay for circulating free and antibody-bound insulin concentrations in insulin-treated diabetics: A re-evaluation of methods. Diabetic Med. 1985;2:456–60.
22. Cresto JC, Lavine RL, Fink G, Recant L. Plasma proinsulin. Comparison of insulin specific protease and gel filtration assays. Diabetes 1974;23:505–11.
23. Damgaard U, Kruse V. Inherent problems in radioimmunoassay exemplified by determination of proinsulin-like immunoreactivity in bovine insulin. In: Hormone Drugs. Proceedings of the FDA-USP Workshop on Drug and Reference Standards for Insulins, Somatropins, and Thyroid-axis Hormones. Bethesda, Maryland, May 19–21, 1982. Rockville, Maryland: United States Pharmacopeial Convention, 1982:187–91.
24. Deacon CF, Conlon JM. Measurement of circulating human proinsulin concentrations using a proinsulin-specific antiserum. Diabetes 1985;34:491–7.
25. Desbuquois B, Aurbach GD. Use of polyethylene glycol to separate free and antibody-bound peptide hormones in radioimmunoassays. J Clin Endocr 1971;33:732–8.
26. Dinesen B, Kappelgaard AM, Hartling S, Binder C, Faber O. High capacity ELISA for human proinsulin. Poster presented at the XII Congress of the International Diabetes Federation, Madrid, Spain, September 23–28, 1985.
27. Duckworth WC, Kitabchi AE, Heinemann M. Direct measurement of plasma proinsulin in normal and diabetic subjects. Am J Med 1972;53:418–27.
28. Ege H, Heding LG. Prolongation and reduction of the effect or circulating insulin caused by insulin antibodies. In: Brunetti P, Alberti KGMM, Albisser AM, Hepp KD, Massi Benedetti M, eds. Artificial Systems for Insulin Delivery. Proceedings of the International Symposium on Artificial Systems for Insulin Delivery, Assisi, Italy, September 20–23, 1981. New York: Raven Press, 1983:69–74.
29. Elkeles RS, Heding LG, Paisey RB. The long-term effects of chlorpropamide on insulin, C-peptide, and proinsulin secretion. Diabetes Care 1982;5:427–9.
30. Elliot RB, Crossley JR, Berryman CC, James AG. Partial preservation of pancreatic B-cell function in children with diabetes. Lancet 1981;2:1–4.
31. Faber OK, Markussen J, Naithani VK, Binder C. Production of antisera to synthetic benzyloxycarbonyl-C-peptide of human proinsulin. Hoppe Seylers Z Physiol Chem 1976;357:751–7.
32. Faber OK, Binder C, Markussen J, et al. Characterization of seven C-peptide antisera. Diabetes 1978;27(Suppl 1):70–7.
33. Faber OK. Beta-cellefunktionen ved insulinbehandlet diabetes mellitus bedømt ved måling af C-peptid. Copenhagen, Denmark: Lægeforeningens Forlag, 1979. 31 pp. Dissertation.
34. Falholt K, Alberti KGMM, Heding LG. Aorta and muscle metabolism in pigs with peripheral hyperinsulinaemia. Diabetologia 1985;28:32–7.
35. Frank BH, Pettee JM, Zimmerman RE, Burck PJ. The production of human proinsulin and its transformation to human insulin and C-peptide. In: Rich DH, Gross E, eds. Peptides: Synthesis – Structure – Function. Proceedings of the Seventh American Peptide Symposium, Madison, Wisconsin, June 14–19, 1981. Rockford: Pierce Chemical Company, 1981:729–38.
36. Garcia-Webb P, Bottomley S, Bonser AM. Instability of C-peptide reactivity in plasma and serum stored at −20°C. Clin Chim Acta 1983; 129:103–6.
37. Gerbitz KD, Kemmler W, Edelmann A, Summer J, Mehnert H, Wieland OH. Free insulin, bound insulin, C-peptide and the metabolic control in juvenile onset diabetics: Comparison of C-peptide secretors and non-secretors during 24 hours conventional insulin therapy. Eur J Clin Invest 1979;9:475–83.
38. Given BD, Cohen RM, Shoelson SE, Frank BH, Rubenstein AH, Tager HS. Biochemical and clinical implications of proinsulin conversion intermediates. J Clin Invest 1985;76:1398–1405.

39. Gorden P, Hendricks CM, Roth J. Circulating proinsulin-like component in man: Increased proportion in hypoinsulinemic states. Diabetologia 1974;10:469–74.
40. Gray IP, Siddle K, Docherty K, Frank BH, Hales CN. Proinsulin in human serum: Problems in measurement and interpretation. Clin Endocrinol 1984;21:43–7.
41. Grodsky GM, Forsham PH. An immunochemical assay of total extractable insulin in man. J Clin Invest 1960;39:1070–9.
42. Grodsky GM. Production of autoantibodies to insulin in man and rabbits. Diabetes 1965;14:396–403.
43. Gunnarsson R, Arner P, Groth CG, Heding LG, Lundgren G, Östman J. Plasma C-peptide as an indicator of human pancreatic graft function. Acta Med Scand (Suppl) 1980;(No 639):55–6.
44. Hales CN, Randle PJ. Immunoassay of insulin with insulin-antibody precipitate. Biochem J 1963;88:137–46.
45. Heding LG. A simplified insulin radioimmunoassay method. In: Donato L, Milhaud G, Sirchis J, eds. Proceedings of the Conference on Problems Connected with the Preparation and Use of Labelled Proteins in Tracer Studies, Pisa, Italy, January 17–19, 1966. Brussels: European Atomic Energy Community, EURATOM, 1966:345–51.
46. Heding LG. Sources of error in radioimmunoassay. In: Donato L, Milhaud G, Sirchis J, eds. Proceedings of the Conference on Problems Connected with the Preparation and Use of Labelled Proteins in Tracer Studies, Pisa, Italy, January 17–19, 1966. Brussels: European Atomic Energy Community, EURATOM, 1966:249–57.
47. Heding LG. Determination of free and antibody-bound insulin in insulin treated diabetic patients. Horm Metab Res 1969;1:145–6.
48. Heding LG. Radioimmunological determination of pancreatic and gut glucagon in plasma. Diabetologia 1971;7:10–9.
49. Heding LG. Immunologic properties of pancreatic glucagon: Antigenicity and antibody characteristics. In: Lefebvre PJ, Unger RH, eds. Glucagon: Molecular Physiology, Clinical and Therapeutic Implications. Oxford, New York: Pergamon Press, 1972:187–200.
50. Heding LG. Determination of total serum insulin (IRI) in insulin-treated diabetic patients. Diabetologia 1972;8:260–6.
51. Heding LG, Larsen UD, Markussen J, Jørgensen KH, Hallund O. Radioimmunoassays for human, pork and ox C-peptides and related substances. Horm Metab Res (Suppl) 1974;5:40–4.
52. Heding LG. Radioimmunological determination of human C-peptide in serum. Diabetologia 1975;11:541–8.
53. Heding LG, Rasmussen SM. Human C-peptide in normal and diabetic subjects. Diabetologia 1975;11:201–6.
54. Heding LG. Radioimmunoassays for C-peptide and proinsulin. In: Bajaj JS, ed. Diabetes. Proceedings of the IX Congress of the International Diabetes Federation, New Delhi, India, October 31–November 5, 1976. Amsterdam. Oxford: Excerpta Medica, 1977:126–33.
55. Heding LG, Specific and direct radioimmunoassay for human proinsulin in serum. Diabetologia 1977;13:467–74.
56. Heding LG, Insulin, C-peptide, and proinsulin in nondiabetics and insulin-treated diabetics. Characterization of the proinsulin in insulin-treated diabetics. Diabetes 1978;27(Suppl 1):178–83.
57. Heding LG, Ludvigsson J. Human proinsulin in insulin-treated juvenile diabetics. Acta Paediatr Scand (Suppl) 1977;(No 270):48–52.
58. Heding LG, Faber O, Kasperska-Czyzykowa T, Sestoft L, Turner R. Radioimmunoassay of proinsulin and hyperproinsulinemic states. In: Baba S, Kaneko T, Yanaihara N, eds. Proinsulin, Insulin, C-peptide. Proceedings of the Symposium on Proinsulin, Insulin and C-peptide, Tokushima, Japan, July 12–14, 1978. Amsterdam, Oxford: Excerpta Medica, 1979:254–61.
59. Heding LG, Kasperska-Czyzykowa T. C-peptide and proinsulin after oral glucose. Acta Med Scand (Suppl) 1980;(No 639):33–6.
60. Heding LG, Persson B, Stangenberg M. β-cell function in newborn infants of diabetic mothers. Diabetologia 1980;19:427–32.
61. Heding LG, Larsson Y, Ludvigsson J. The immunogenicity of insulin preparation. Anti-

body levels before and after transfer to highly purified porcine insulin. Diabetologia 1980;19:511–5.

62. Heding LG, Proinsulin, glucagon, and pancreatic polypeptide radioimmunoassays. In: Gueriguian JL, ed. Insulins, Growth Hormone and Recombinant DNA Technology. New York: Raven Press, 1981:87–97.
63. Heding LG, Ludvigsson J, Kasperska-Czyzykowa T. β-cell secretion in non-diabetics and insulin-dependent diabetics. Acta Med Scand (Suppl) 1981;(No 656):5–9.
64. Heding LG, The immunogenicity of glucagon. In: Lefèbvre PJ, ed. Glucagon I. Handbook of Experimental Pharmacology, Vol. 66/I. Berlin, Heidelberg, New York, Tokyo: Springer-Verlag, 1983:189–201.
65. Heding LG, Kruse V. Usefulness of fasting proinsulin in the diagnosis of insulinoma. Diabetes 1984;33(Suppl 1):148.
66. Heding LG, Marshall MO, Persson B, et al. Immunogenicity of monocomponent human and porcine insulin in newly diagnosed Type 1 (insulin-dependent) diabetic children. Diabetologia 1984;27:96–8.
67. Hoekstra JBL, van Rijn HJM, Ërkelens DW, Thijssen HH. C-peptide. Diabetes Care 1982;5:438–46.
68. Horwitz DL, Starr JI, Mako ME, Blackard WG, Rubenstein AH. Proinsulin, insulin, and C-peptide concentrations in human portal and peripheral blood. J Clin Invest 1975;55:1278–83.
69. Jørgensen KH, Binder C. ¹²⁵I-insulin as a tracer of insulin in different chemical processes. In: Donato L, Milhaud G. Sirchis J, eds. Proceedings of the Conference on Problems Connected with the Preparation and Use of Labelled Proteins in Tracer Studies, Pisa, Italy, January 17–19, 1966. Brussels: European Atomic Energy Community, EURATOM, 1966:329–32.
70. Jørgensen KH, Larsen UD. Homogeneous mono-¹²⁵I-insulins. Preparation and characterization of mono-¹²⁵I-(Tyr A14)- and mono-¹²⁵I-(Tyr A19)-insulin. Diabetologia 1980;19:546–54.
71. Jørgensen KH. Aspects of insulin immunogenicity. In: Sakamoto N, Alberti KGMM, Hotta N, eds. Recent trends in management of diabetes mellitus. Proceedings of the Second International Symposium on Treatment of Diabetes Mellitus, Nagoya, Japan, November 13–15, 1985. Amsterdam, New York, Oxford: Excerpta Medica, 1987:111–22.
72. Kaneko T, Munemura M, Oka H, et al. Demonstration of C-peptide immunoreactivity in various body fluids and clinical evaluation of the determination of urinary C-peptide immunoreactivity. Endocrinol Jpn 1975;22:207–12.
73. Kasperska-Czyzykowa T, Heding LG, Czyzyk A. Serum levels of true insulin, C-peptide and proinsulin in peripheral blood in patients with cirrhosis. Diabetologia 1983;25:506–9.
74. Kato K, Umeda Y, Suzuki F, Hayashi D, Kosaka A. Evaluation of a solid-phase enzyme immunoassay for insulin in human serum. Clin Chem 1979;25:1306–8.
75. Kruse V. Effect of insulin-binding antibodies on free insulin in plasma and tissue after subcutaneous injection. A model study. In: Keck K, Erb P, eds. Basic and Clinical Aspects of Immunity to Insulin. Proceedings of the International Workshop, Konstanz, Germany, September 28–October 1, 1980. Berlin, New York: Walter de Gruyter, 1980:319–34.
76. Kruse V, Heding LG, Jørgensen KH, et al. Human proinsulin standards. Diabetologia 1984;27:414–5.
77. Kuzuya T, Matsuda A, Saito T, Yoshida S. Human C-peptide immunoreactivity (CPR) in blood and urine. Evaluation of a radioimmunoassay method and its clinical applications. Diabetologia 1976;12:511–8.
78. Kuzuya H, Blix PM, Horwitz DL, et al. Heterogeneity of circulating C-peptide. J Clin Endocrinol Metab 1977;44:952–62.
79. Kuzuya H, Blix PM, Horwitz DL, Steiner DF, Rubenstein AH. Determination of free and total insulin and C-peptide in insulin-treated diabetics. Diabetes 1977;26:22–9.
80. Kuzuya H, Blix PM, Horwitz DL, et al. Heterogeneity of circulating human C-peptide. Diabetes 1978;27(Suppl 1):184–91.
81. Kuzuya H, Chance RE, Steiner DF, Rubenstein AH. On the preparation and characterization of standard materials for natural human proinsulin and C-peptide. Diabetes 1978;27(Suppl 1):161–9.

82. Kühl C, Faber OK, Hornnes P, Jensen SL. C-peptide metabolism and the liver. Diabetes 1978;27(Suppl 1):197–200.
83. Lambert B, Jacquemin C. About insulin iodination. I. Monoiodination of insulin by horseradish peroxidase. Biochimie 1973;55:1395–1400.
84. Ludvigsson J, Heding LG. C-peptide in children with juvenile diabetes. Diabetologia 1976;12:627–30.
85. Ludvigsson J, Heding LG. Beta-cell function in children with diabetes. Diabetes 1978;27(Suppl 1):230–4.
86. Ludvigsson J, Heding LG, Lernmark Å, Lieden G. An attempt to break the autoimmune process at the onset of IDDM by the use of plasmapheresis or high doses of prednisolone. International Study Group on Diabetes in Children and Adolescents Bulletin 1982;(No 6):11–2.
87. Ludvigsson J, Heding L. Abnormal proinsulin/C-peptide ratio in juvenile diabetes. Acta Diabetol Lat 1982;19:351–8.
88. Ludvigsson J, Heding L, Lieden G, Marner B, Lernmark Å. Plasmapheresis in the initial treatment of insulin-dependent diabetes mellitus in children. Br Med J 1983;286:176–8.
89. Madsbad S, Alberti KGMM, Binder C, et al. Role of residual insulin secretion in protecting against ketoacidosis in insulin-dependent diabetes. Br Med J 1979;2:1257–9.
90. Madsbad S, Krarup T, McNair P, et al. Practical clinical value of the C-peptide response to glucagon stimulation in the choice of treatment in diabetes mellitus. Acta Med Scand 1981;210:153–6.
91. Madsbad S, Krarup T, Faber OK, Binder C, Regeur L. The transient effect of strict glycaemic control on B cell function in newly diagnosed type 1 (insulin-dependent) diabetic patients. Diabetologia 1982;22:16–20.
92. Madsbad S. Prevalence of residual B cell function and its metabolic consequences in type 1 (insulin-dependent) diabetes. Diabetologia 1983; 24:141–7.
93. Madsbad S, Hilsted J, Krarup T, Sestoft L, Christensen NJ, Tronier B. The importance of plasma free insulin and counterregulatory hormones for the recovery of blood glucose following hypoglycaemia in Type 1 diabetics. Acta Endocrinol (Copenh) 1985;108:224–30.
94. Madsen OD, Frank BH, Steiner DF. Human proinsulin-specific antigenic determinants identified by monoclonal antibodies. Diabetes 1984;33:1012–6.
95. Markussen J, Jørgensen KH, Heding LG. Preparation of bovine ^{125}I-tyrosyl-C-peptide. Horm Metab Res 1970;2:53–5.
96. Markussen J, Sundby F, Smyth DG, Ko A. Preparation of human C-peptide. Horm Metab Res 1971;3:229–32.
97. Markussen J, Damgaard U, Pingel M, Snel L, Sørensen AR, Sørensen E. Human insulin (NOVO): Chemistry and characteristics. Diabetes Care 1983;6(Suppl 1):4–8.
98. Markussen J, Damgaard U, Diers I, et al. Biosynthesis of human insulin in yeast via single-chain precursors. In: Theodoropoulus D, ed. Peptides 1986. Proceedings of the 19th European Peptide Symposium, Porta Carras, Chalkidiki, Greece, August 31–September 5, 1986. Berlin, New York: Walter de Gruyter, 1987:189–94.
99. Meade RC, Klitgaard HM. A simplified method for immuno-assay of human serum insulin. J Nucl Med 1962;3:407–16.
100. Melani F, Steiner DF. Proinsulin and C-peptide in human serum. Acta Diabetol Lat 1970;7(Suppl 1):107–21.
101. Mirsky A, Kawamura K, Riley EJ. Heterogeneity of crystalline insulin. Endocrinology 1966;78:1115–9.
102. Morgan CR, Lazarow A. Immunoassay of insulin: Two antibody system. Plasma insulin levels of normal, subdiabetic and diabetic rats. Diabetes 1963;12:115–26.
103. Naithani VK, Dechesne M, Markussen J, Heding LG. Studies on polypeptides, V. Improved synthesis of human proinsulin C-peptide and its benzyloxycarbonyl derivative. Circular dichroism and immunological studies of human C-peptide. Hoppe Seylers Z Physiol Chem 1975;356:997–1010.
104. Naithani VK, Dechesne M, Markussen J, Heding LG, Larsen UD. Studies on polypeptides, VI. Synthesis, circular dichroism and immunological studies of tyrosyl C-peptide of human proinsulin. Hoppe Seylers Z Physiol Chem 1975;356:1305–12.
105. Nakagawa S, Nakayama H, Sasaki T, et al. A simple method for the determination of serum free insulin levels in insulin-treated patients. Diabetes 1973;22:590–600.

106. Nerup J, Bendtzen K, Mandrup-Poulsen T, A role for cyclosporin A in the treatment of insulin-dependent diabetes mellitus? Diabetic Med 1985;2:441-6.
107. Nicol DSHW, Smith LF. Amino-acid sequence of human insulin. Nature 1960;187:483-5.
108. Ohneda A, Toyota T, Sato S, Yamagata S. Extraction of plasma insulin in radioimmunoassay for removal of nonspecific inhibitor and of circulating insulin antibody. Tokushima J Exp Med 1970;100:75-84.
109. Owens DR, Jones MK, Birtwell AJ, et al. Pharmacokinetics of subcutaneously administered human, porcine and bovine neutral soluble insulin to normal man. Horm Metab Res 1984;16(Suppl 1):195-9.
110. Pearson MJ, Martin FIR. The separation of total plasma insulin from binding proteins using gel filtration: Its application to the measurement of rate of insulin disappearance. Diabetologia 1970;6:581-5.
111. Persson B, Heding LG, Lunell NO, Pschera H, Stangenberg M, Wager J. Fetal beta cell function in diabetic pregnancy. Am J Obstet Gynecol 1982;144:455-9.
112. Pingel M, Vølund AA, Sørensen E, Sørensen AR. Assessment of insulin potency by chemical and biological methods. In: Hormone Drugs. Proceedings of the FDA-USP Workshop on Drug and Reference Standards for Insulins, Somatropins, and Thyroid-axis Hormones, Bethesda, Maryland, May 19-21, 1982. Rockville, Maryland: United States Pharmacopeial Convention, 1982:200-7.
113. Polonsky K, Jaspan J, Pugh W, et al. Metabolism of C-peptide in the dog. In vivo demonstration of the absence of hepatic extraction. J Clin Invest 1983;72:1114-23.
114. Rainbow SJ, Woodhead JS, Yue DK, Luzio SD, Hales CN. Measurement of human proinsulin by an indirect two-site immunoradiometric assay. Diabetologia 1979;17:229-34.
115. Rehfeld JF, Stadil F, Rubin B. Production and evaluation of antibodies for the radioimmunoassay of gastrin. Scand J Clin Lab Invest 1972;30:221-32.
116. Roth J, Gorden P, Pastan I. 'Big insulin': A new component of plasma insulin detected by immunoassay. Proc Natl Acad Sci USA 1968;61:138-45.
117. Rubenstein AH, Cho S, Steiner DF. Evidence for proinsulin in human urine and serum. Lancet 1968;1:1353-5.
118. Rubenstein AH, Steiner DF. Human proinsulin: Some considerations in the development of a specific immunoassay. In: Camerini-Dávalos RA, Cole HS, eds. Early Diabetes. New York, London: Academic Press, 1970:159-66.
119. Rubenstein AH, Block MB, Starr J, Melani F, Steiner DF, Proinsulin and C-peptide in blood. Diabetes 1972;21(Suppl 2):661-72.
120. Schlichtkrull J, Brange J, Ege H. et al. Proinsulin and related proteins. Diabetologia 1970;6:80-1.
121. Schlichtkrull J, Brange J, Christiansen AH, Hallund O, Heding LG, Jørgensen KH. Clinical aspects of insulin. Antigenicity. Diabetes 1972;21(Suppl 2):649-56.
122. Schlichtkrull J, Brange J, Christiansen AH, et al. Monocomponent insulin and its clinical implications. Horm Metab Res (Suppl) 1974;5:134-43.
123. Schlichtkrull J. Insulin in perspective. IDF Bulletin 1979;24:7-10.
124. Sestoft L, Heding LG. Hypersecretion of proinsulin in thyrotoxicosis. Diabetologia 1981;21:103-7.
125. Skom JH, Talmage DW. Nonprecipitating insulin antibodies. J Clin Invest 1958;37:783-6.
126. Steiner DF, Oyer PE. The biosynthesis of insulin and a probable precursor of insulin by a human islet cell adenoma. Proc Natl Acad Sci USA 1967;57:473-80.
127. Steiner DF, Cunningham D, Spigelman L, Aten B. Insulin biosynthesis: Evidence for a precursor. Science 1967;157:697-700.
128. Steiner DF. Evidence for a precursor in the biosynthesis of insulin. NY Acad Sci 1967;30:60-8.
129. Steiner DF, Hallund O, Rubenstein A, Cho S, Bayliss C. Isolation and properties of proinsulin intermediate forms, and other minor components from crystalline bovine insulin. Diabetes 1968;17:725-36.
130. Steiner DF. On the role of the proinsulin C-peptide. Diabetes 1978; 27(Suppl 1):145-8.
131. Stiller CR, Laupacis A, Dupre J, et al. Cyclosporine for treatment of early type 1 diabetes: Preliminary results. N Engl J Med 1983;308:1226-7.

132. Stoll RW, Touber JL, Ensinck JW, Williams RH. Substances immunologically related to proinsulin or connecting peptide in swine plasma. Horm Metab Res 1970;2:153–6.
133. Sundby F, Markussen J. Preparative method for the isolation of C-peptides from ox and pork pancreas. Horm Metab Res 1970;2:17–20.
134. Sutcliffe N, Bristow AF. The proinsulin content of commercial bovine insulin formulations. J Pharm Pharmacol 1984;36:163–6.
135. Track NS, Frerichs H, Creutzfeldt W. Release of newly synthesized proinsulin and insulin from granulated and degranulated isolated rat pancreatic islets. The effect of high glucose concentration. Horm Metab Res (Suppl) 1974;5:97–103.
136. Tronier B, Heding LG, Kruse V. C-peptide usefulness in diabetes research. C-peptide determination, methodological aspects. In: Cabezas-Cerrato J, ed. U-100 Peptido C. Proceedings of the III Symposium Satelite de Novo al VII Congreso Nacional de la Sociedad Española de Diabetes, Santiago de Compostela, Spain, June 6, 1984. Madrid: Jarpyo Editories, S.A., 1984;11–20. (available from Novo Industri A/S, Novo Alle, DK-2880 Bagsværd, Denmark).
137. Turner RC, Heding LG. Plasma proinsulin, C-peptide and insulin in diagnostic suppression tests for insulinomas. Diabetologia 1977;13:571–7.
138. Wu CY, Jørgensen PN, Patkar SA, Kruse V, Heding LG, Zeuthen J. Characterization of monoclonal antibodies against bovine insulin. Ann Immunol (Paris) 1986;C137:11–23.
139. Yalow RS, Berson SA. Immunoassay of endogenous plasma insulin in man. J Clin Invest 1960;39:1157–75.
140. Yalow RD, Berson SA. Immunological specificity of human insulin: Application to immunoassay of insulin. J Clin Invest 1961;40:2190–8.
141. Yanaihara N, Yanaihara C, Sakagami M, Sakura N, Hashimoto T, Nishida T. Syntheses of C-peptides and human proinsulin. Diabetes 1978;27(Suppl 1):149–60.
142. Yip CC, Logothetopoulos J. A specific anti-proinsulin serum and the presence of proinsulin in calf serum. Proc. Natl Acad Sci USA 1969; 62:415–9.
143. Ørskov H. Wick-chromatography for the immunoassay of insulin. Scand J Clin Lab Invest 1967;20:297–304.
144. Östman J, Arner P, Groth CG, Gunnarsson R, Heding LG, Lundgren G. Plasma C-peptide and serum insulin antibodies in diabetic patients receiving pancreatic transplants. Diabetologia 1980;19:25–30.

Diabetologia 8, 260—266 (1972)
© by Springer-Verlag 1972

Determination of Total Serum Insulin (IRI) in Insulin-treated Diabetic Patients

Lise G. Heding

Novo Research Institute, Copenhagen, Denmark

Received: December 28, 1971, accepted: March 20, 1972

Summary. A routine method is described for the determination of total IRI (immunoreactive insulin) in insulin-treated diabetics. The method involves an easy acid ethanol extraction, whereby antibody-bound IRI is dissociated and separated, together with the "free" IRI from the serum proteins and the antibodies. The recovery of IRI in this procedure is about 80%. After the separation, the isolated total IRI is measured in an immunoassay, using ethanol for the separation of free and antibody bound ^{125}I-insulin. In 169 diabetic patients treated with insulin in doses of from 6 to 120 units/day, the fasting serum total IRI was between 6 and 4374 μU/ml, with a mean of 392 μU/ml. During treatment with insulin, the level of total IRI increased from normal values, registered during the first two months, to a higher level which became stable after about 5 months of treatment. The increase in IRI occurred simultaneously with the formation of antibodies. Insulin-resistant patients showed very high IRI levels.

Determination de l'insuline totale chez les diabétiques traités a l'insuline

Résumé. On décrit une méthode de routine pour le dosage de l'IRI (insuline immunoréactive) totale chez les diabétiques traités par l'insuline. La méthode comprend une extraction à l'acide-éthanol, très simple, pendant laquelle l'IRI liée aux anticorps est dissociée et séparée ainsi que l'IRI »libre« des protéines sériques, anticorps compris. La récupération de l'IRI par cette méthode est aux environs de 80%. Après la séparation, l'IRI totale isolée est mesurée par un dosage immunologique qui se sert de l'éthanol afin de séparer l'I^{125}-insuline libre de celle liée aux anticorps. Chez 169 malades diabétiques traités par l'insuline à des doses allant de 6 à 120 unités par jour, l'IRI totale sérique à jeun était de 6 à 4374 μU/ml, avec une moyenne de 392 μU/ml. Pendant le traitement par

l'insuline le taux de l'IRI totale est passé des niveaux normaux, enregistrés pendant les deux premiers mois, à des niveaux plus éléévs qui se stabilisent 5 mois environ apres le début du traitement. L'augmentation de l'IRI coïncide avec la formation d'anticorps. Les malades insulino-résistants présentent des valeurs très hautes d'IRI.

Bestimmung des Gesamtserum-Insulins (IRI) bei insulinbehandelten Diabetikern

Zusammenfassung. Für die Bestimmung des Gesamt-IRI (immunoreaktiven Insulins) bei Diabetikern, die mit Insulin behandelt wurden, wird eine Routinemethode beschrieben. Die Methode schließt eine einfache Säure-Äthanol-Extraktion ein, wobei das antikörpergebundene IRI dissoziiert und zusammen mit dem „freien" IRI von den Serumproteinen, einschließlich den Antikörpern, getrennt wird. Bei diesem Verfahren werden etwa 80% des IRI wiedergefunden. Nach der Trennung wird das isolierte Gesamt-IRI immunologisch gemessen. Für die Trennung des freien von dem an Antikörper gebundenen ^{125}I-Insulin wird Äthanol verwendet. Bei 169 Diabetikern, die mit 6 bis 120 E Insulin/Tag behandelt wurden, lag das Nüchternserum-Gesamt-IRI zwischen 6 und 4374 μE/ml (Mittelwert 392 μE/ml). Im Laufe der Insulinbehandlung stieg das Gesamt-IRI von Normalwerten, die während der ersten 2 Monate registriert wurden, auf ein höheres Niveau an, das sich nach etwa 5 Monaten Behandlungsdauer stabilisierte. Der Anstieg des IRI erfolgte gleichzeitig mit der Bildung von Antikörpern. Bei insulinresistenten Patienten ergaben sich sehr hohe IRI-Konzentrationen.

Key words: Insulin, radioimmunoassay, total IRI in insulin-treated diabetics, acid ethanol extraction of insulin.

Introduction

Yalow and Berson (1960) were the first to develop and describe the insulin radioimmunoassay, and since then this assay has been widely used, either in its original form or in a form modified, e.g., with respect to the tracer and/or the separation technique, in order to meet the requirements of a routine method. As a result of the availability of these routine methods, an enormous amount of information has been collected over the past 11 years about the concentration of immunoreactive insulin (IRI) in serum from normal persons and untreated diabetic patients. However, information about IRI levels in insulin-treated patients is wanting.

When diabetic patients are treated with commercial insulin preparations, nearly all of them develop

insulin antibodies (Berson and Yalow, 1964) after a few months of treatment. The serum from these patients then contains a mixture of „free" insulin, antibody-bound insulin and free antibodies. Due to the presence of antibodies, it is not possible to determine the total amount of IRI by a direct immunoassay such as those used for the sera from normals and diabetics not treated with insulin. The antibodies must be removed from the diabetic serum before an IRI determination can be performed.

Grodsky (1965) extracted insulin from the sera of two resistant patients and found values as high as 12 mU of total insulin per ml of serum. Heding and Vølund (1967) and Heding (1969) described a routine method of acid extraction for the determination of total insulin in insulin-treated patients. In a series of insulin-treated diabetic patients (10 to 120 i. U. a day) the IRI values were found to range between 0 and 3000 µU/ml. Ohneda *et al.* (1970) described a plasma extraction method which abolished a non-specific inhibitor of the double-antibody method. Furthermore, the extraction removed the antibodies present in the serum, making it possible to determine IRI in sera from insulin-treated diabetics. Determination of IRI performed in eight sera gave values of from 10 to 390 µU/ml. Pearson and Martin (1970) developed gel filtration, on Sephadex G-50, of plasma from diabetic patients after dissociation of the insulin-antibody complex at low pH. Using this method, the authors found between 700 and 6000 µU IRI per ml of plasma of six fasting, insulin-treated diabetic patients.

This paper describes a routine method for the determination of total IRI in insulin-treated patients, and consists of two parts: 1. insulin radioimmuno-assay and 2. determination of total IRI in serum from insulin-treated persons.

Isolation of total IRI from serum

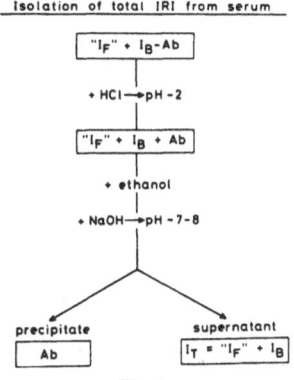

Fig. 1

1. *Insulin radioimmunoassay.* The general principles of radioimmunoassay are well known. One of the most critical steps in this assay is the separation of free and antibody-bound ^{125}I-insulin. A number of techniques are available today (Kirkham and Hunter, 1971).

In the ethanol method, the free and the antibody-bound insulin are separated by addition of 96% ethanol to give a final concentration of 79%. This method was developed (Heding, 1966) as a substitute for the paper chromatographic method and the double-antibody method, to avoid the time consumed by the former and the sources of error of the latter.

2. *Determination of total IRI in serum of insulin-treated persons.* The principle of this method (Heding, 1967, 1969) is shown in Fig. 1. The serum sample contains free IRI, antibody-bound IRI and a surplus of antibodies. The pH is adjusted to approximately 2 with HCl; at this low pH the insulin-antibody complex

dissociates immediately into free insulin and anti-body. Ethanol is then added, and due to the low pH no precipitation of the serum proteins will take place until NaOH is added to bring the pH close to 7. The ethanol concentration of this neutral mixture is approximately 75%, at which level insulin does not react with antibody. The insulin remains in the super-natant and is separated from the precipitated anti-bodies and other serum proteins by centrifugation. The supernatant is evaporated in an exsiccator in vacuum, and the residue is dissolved in buffer and immunoassayed.

Materials and Methods

Monocomponent-insulins of human, porcine and bovine origin, with biological activities of 25.4, 27.2 and 25.7 i.U./mg, respectively, (supplied by the NOVO Insulin Laboratory), were used as standards.

^{125}I-pork insulin was used as tracer. It was prepared by iodinating monocomponent-pork insulin with ^{125}I according to Jørgensen and Binder (1966). Its specific activity was approximately 30 mCi/mg. ^{125}I-insulin and anti-insulin sera were diluted in phosphate buffer (0.04 M, pH 7.4) containing human albumin (Behringwerke) (1 g/l) and thiomersal (0.2 g/l) (subsequently referred to as FAM). All standards and samples were dissolved and diluted in phosphate buffer (0.04 M, pH 7.4) containing NaCl (6 g/l), human albumin (60 g/l) and thiomersal (0.2 g/l) (subsequently referred to as NaFAM).

Insulin antibodies were raised by injecting guinea pigs (weighing 300—400 g) with 0.5 ml of an emulsion of 3.75 ml sterile water and 6.25 ml of Zinc Protamine Insulin (NOVO, pork insulin, 80 i.U./ml) and 10 ml of Freund's adjuvant, corresponding to a dose of 12.5 units of insulin; this dose was subsequently increased to 25 units. The guinea pigs were allowed 10% glucose water *ad libitum* after the injections.

The radioimmunoassay procedure was as follows: to triplicates of 100 µl of standard solutions (containing from 10 to 100 µU of insulin/ml) or samples was added 100 µl of anti-insulin guinea-pig serum diluted 1 : 35000. After 20 h of incubation at 4°C, 100 µl of ^{125}I-insulin (200 µU/ml) was added at 4°C, and after another incubation period at 4°C (4—20 h), 1.6 ml of ethanol was added in order to separate the free and the antibody-bound insulin. After mixing and centrifugation for 10 min at 2500 rpm, the supernatant containing the free ^{125}I-insulin was decanted into disposable plastic tubes (diameter: 10 mm, height: 75 mm, NUNC, Roskilde, Denmark), the tubes were stoppered with plastic stoppers, and the radio-activity counted.

Blood was drawn from the antecubital vein into glass tubes and allowed to clot for one hour at room temperature before centrifugation. The serum was pipetted into plastic tubes and stored at −18°C until used.

Determination of the total amount of IRI was done as follows: to duplicates of 500 µl of serum was added 100 µl of 1 N HCl to give a pH of approximately 2.5. The tubes were shaken and incubated at room temperature for 10 min, whereupon 2.5 ml of 95% ethanol was added and the contents mixed by vigorously inverting the tubes. Due to the low pH, the plasma proteins did not precipitate. 100 µl of 1 N NaOH was then added, and again, the mixture was shaken vigorously. A heavy protein precipitate was formed. The precipitate was separated by centrifugation for 10 min at about 2000 G, and the supernatant, containing the insulin, was transferred to a small bottle and evaporated to dryness in a dessiccator under a pressure of 10 mm Hg overnight. One hundred samples could easily be handled simultaneously. The dry residue was dissolved in 1 ml of NaFAM buffer and immunoassayed.

Results

1. Insulin radioimmunoassay

Concentration of ethanol, salt, albumin

Two requirements should be fulfilled in connection with the use of ethanol for the separation of free and antibody-bound ^{125}I-insulin, viz.: 1. complete precipitation of all the ^{125}I-insulin in an incubation medium containing insulin antibodies in surplus, and 2. no precipitation of ^{125}I-insulin in solutions containing no antibodies. As previously shown (Heding, 1966), these requirements are fulfilled by adding 1.6 ml of 96% ethanol to 0.3 ml of an incubation mixture to give a final ethanol concentration of approximately 79%. Complete immunoassays (including standard curves and serum samples) were run simultaneously using double-antibody, paper chromatographic or ethanol separation techniques. No significant differences could be found in the per cent ^{125}I-insulin bound to antibodies either in the standards or in the samples.

A minimum concentration of 0.02 M Cl$^-$ (or other monovalent anions) was found necessary to ensure complete precipitation of the insulin-antibody complex; furthermore, the insulin-antibody complex was shown to be stable in 79% ethanol for several hours.

The protein concentration influences the coprecipitation of the ^{125}I-insulin. Coprecipitation of free ^{125}I-insulin in the presence of different serum samples and 6% albumin containing buffer was between 4.6% and 5.2%. It is important that the protein concentration should be the same in all incubation mixtures, and a final concentration of 2% has been used throughout.

At temperatures below 10°C, the coprecipitation increases, and the temperature in the final mixture should therefore be above 10°C (e.g., ethanol at room temperature).

Adsorption of ^{125}I-insulin to different types of tubes, concentration of ^{125}I-insulin in foam. It is a well-known phenomenon that insulin in low concentration (ng range) is adsorbed to the surface of glass-ware unless other substances are present — preferably proteins. Probably, these proteins, e.g. albumin, compete with the insulin for the binding site on the glass-ware, and if albumin is present in great excess, the loss of insulin due to adsorption will be minimal. However, the degree of insulin adsorption depends not only on the amount of protein present but also on the type of glass or plastic-ware used. The adsorption of ^{125}I-labelled ox proinsulin to four different types of glass and plastic tubes was examined. It was found that the plastic tube absorbed 2% of the ^{125}I-labelled ox proinsulin, whereas the glass tubes adsorbed between 13 and 37% of the tracer at 0.1% albumin concentration.

The adsorption was diminished by addition of, e.g., serum or by increasing the concentration of albumin. The adsorption to glass-ware was also demonstrated using ^{125}I-insulin. It is obvious that the type of tube used for immunoassay should be carefully selected. The same holds true for the glass-ware that is used to store and prepare the ^{125}I-insulin solutions.

Solutions of ^{125}I-insulin in FAM containing 0.1% albumin foam readily upon shaking. The concentration of ^{125}I-insulin is higher in the foam than in the solution. This circumstance may induce serious errors. A frozen solution of ^{125}I-insulin was shaken during thawing and 100 μl volumes were pipetted into tubes and counted. The first 90 tubes contained the same amount of ^{125}I-insulin. In the subsequent tubes the radioactivity increased, with the final 4—5 tubes exhibiting very high concentrations of ^{125}I-insulin. Obviously, the foam with its higher concentration of ^{125}I-insulin had gradually settled thereby increasing the concentration of ^{125}I-insulin in the solution.

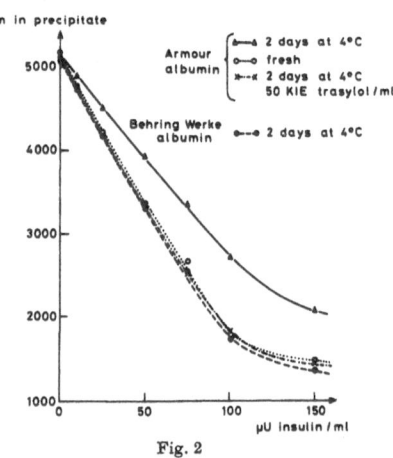

Degradation of insulin by different types of albumin

cpm in precipitate

Armour albumin ⟨ 2 days at 4°C / fresh / 2 days at 4°C 50 KIE trasylol/ml ⟩

Behring Werke albumin — 2 days at 4°C

μU insulin/ml

Fig. 2

Quality of albumin and ^{125}I-insulin; degradation of insulin by serum. As mentioned before, albumin is added to all buffer solutions in order to prevent adsorption of insulin to glass-ware and to attain the same protein concentration in serum and standards. Many publications mention using bovine albumin fraction V (Armour) for this purpose. It was found, however, that this albumin contained enzymes capable of degrading insulin. The problem is illustrated in Fig. 2. A series of insulin standard solutions was prepared with bovine albumin made by Armour, and another one with human albumin made by Behringwerke. It was found that the insulin solution prepared with bovine albumin retained only about 65% of its original content of insulin after 2 days of storage at 4°C. The reason was, no doubt, enzymatic degradation (and not the binding of insulin to albumin) because this degradation could be prevented with trasylol (a proteinase inhibitor).

A common "must" in all immunoassays is the use of a high-quality tracer. The significance of this requirement varies with the different methods of separation. Table 1 shows some results obtained with two

Table 1. *Different qualities of* [125]*I-insulin, effect of binding to antibody and coprecipitation*

| Incubation mixture | % precipitated | |
	^{125}I-insulin[a] K. Jørgensen	^{125}I-insulin[a] commercial
surplus of antibody ^{125}I-insulin	97.6	84
N-serum I ^{125}I-insulin	4.3	25
N-serum II ^{125}I-insulin	4.5	23
NaFAM, 6% albumin ^{125}I-insulin	5.0	23

[a] the concentration of ^{125}I-insulin in both preparations was approx. 220 µU/ml.

different ^{125}I-insulin preparations: one of them was labelled by the method of K. Jørgensen (Jørgensen and Binder, 1966), the other one was a commercially available preparation. It is obvious that a part of the radioactivity in the commercial preparation was not bound to insulin since only 84% could be "bound" to insulin antibodies in surplus, and since more than 20% was precipitated in the absence of antibody.

The degradation of insulin in plasma and serum has been discussed in numerous papers. No detectable degradation occurs in serum stored at −18°C; the problem first arises when serum is handled at temperatures above 0°C. No difference could be shown in the IRI content of six serum samples obtained from a normal person during an oral glucose tolerance test after 4 h of storage at −18°C, 4°C and 30°C (IRI values between 17 and 68 µU/ml).

Reproducibility of standard curves, standard deviation of triplicates, day-to-day variation in serum IRI determinations, normal IRI values. Fig. 3 shows the reproducibility of 7 different standard curves obtained over a period of three weeks. The day-to-day variation is small, but it is advisable always to set up a standard curve together with the samples to be assayed.

The counts from the immunoassay were recorded direct on a punch tape, which was transferred to an IBM 1130 computer and processed according to a program developed by Vølund (1972). The results were written as mean values with the 95% interval. In the range of 0−80 µU/ml, the 95% interval of 59 determinations (2 experiments) was 2.83 µU/ml ± 0.82 (mean ± s.d.). The 95% interval was the same

throughout the entire range, meaning that the most accurate determinations are obtained using the upper part of the standard curve. Two normal sera were assayed in 10 different immunoassays, giving the following results: 39.6 ± 3.0 and 81.1 ± 4.8 µU/ml (mean ± s.d.). Dilution of the serum samples gave the expected values, and recovery of added insulin was practically 100%. Twenty-four normal fasting persons showed IRI values between 0 and 16 µU/ml (mean 7.2 µU/ml). One hour after 1.75 g of glucose/kg, their IRI values were between 30 and 150 µU/ml (mean 69.6 µU/ml).

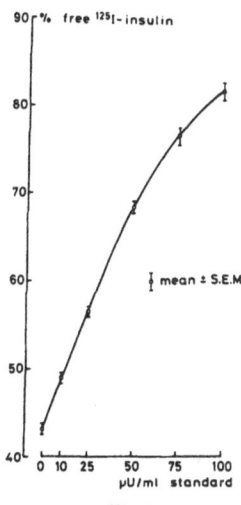

Fig. 3

2. Determination of total IRI in serum from insulin-treated persons

Dissociation of the insulin-antibody complex. A prerequisite of a reliable determination of the total IRI content in antibody-containing diabetic sera is a complete dissociation of the insulin-antibody complex and subsequent removal of the antibodies. Fig. 4 shows the dissociation of a ^{125}I-insulin-antibody complex after addition of HCl to a pH of approximately 2.5. Serum from an insulin treated diabetic patient was incubated with ^{125}I-insulin for 24 h at 4°C, after which 67% of the labelled insulin was bound. HCl was added to give a pH of approximately 2.5. The amount of bound ^{125}I-insulin was determined at various intervals after the addition of HCl by adding ethanol. It was found that the dissociation of the ^{125}I-insulin-antibody complex was immediate and complete. Only 3.5% of the ^{125}I-insulin was precipitated together with the proteins.

Recovery; serial dilutions of the extracted insulin.
The recovery of insulin in the extraction procedure
has been determined using the following three approaches:
1. addition of ^{125}I-insulin to normal and diabetic sera
2. use of normal sera with a known insulin concentration
3. use of normal sera with a known insulin concentration added to guinea-pig anti-insulin serum (AIS).

The recovery of ^{125}I-pork insulin added to diabetic
sera varied from 79 to 84% in 12 sera. The recovery of
insulin from standard solutions containing from 25 to

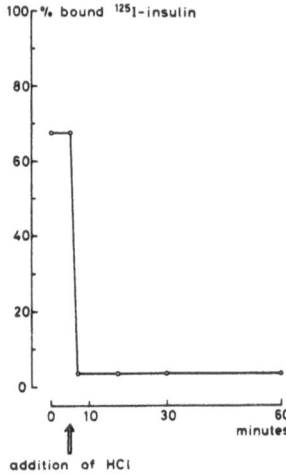

Fig. 4

150 μU/ml was from 72 to 88% (mean 77%). The
recovery of insulin from a normal serum with and
without addition of a surplus of AIS was about 75%
in both cases.

The day-to-day variation in extraction yield was
determined in regard to 3 normal sera with different
levels of insulin. The sera contained 150, 90 and
41 μU/ml and the mean recovery $\pm$ s.d. was $76 \pm 2\%$
(9 extractions), $76 \pm 5\%$ (14 extractions) and $68 \pm$
10%, respectively. As expected, the recovery was
slightly lower and exhibited more variation at low
insulin concentrations. Twenty-eight triplicate extractions carried out the same day, with the extracts
assayed simultaneously, showed a mean s.d. of 2.6%.

Insulin extracted from human sera (normal as well
as that of insulin-treated diabetics) was dissolved in
buffer and serial dilutions were prepared and immunoassayed. A linear dilution curve was obtained in all
cases. Three examples are given in Fig. 5. The extracts
were also checked for their contents of insulin antibodies and they were found not to contain any.

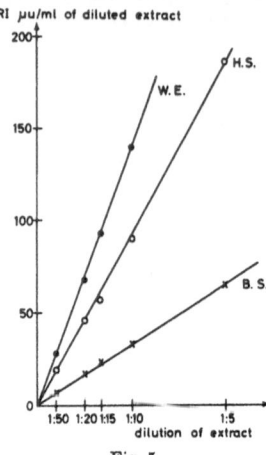

Fig. 5

Table 2. *Total insulin in insulin-treated diabetic patients*

Daily insulin dose i.U.	No. of patients	mean	Total IRI μU/ml range	s.e.m.
6—20	20	252	6—1192	75
21—40	110	341	8—2060	49
41—60	35	543	32—4347	147
61—80	3	492	48—1250	380
>80	1	3000	—	—

Table 3. *Determination of total fasting IRI in two patients
before and after treatment with insulin*

Patient	Date	Total IRI	Daily insulin dose
H.B.	16. 6. 67	24	0
	8. 8. 67	22	8 i.U. Lente
	25. 8. 67	13	8 i.U. Lente
	25. 9. 67	178	14 i.U. Lente
	11. 12. 67	2221	20 i.U. Lente
	19. 12. 67	2172	28 i.U. Lente
	11. 3. 68	1895	36 i.U. Lente
	22. 3. 68	1815	40 i.U. Lente
E.M.	20. 7. 67	14	0
	31. 8. 67	34	12 i.U. Lente
	6. 10. 67	109	10 i.U. Lente
	1. 12. 67	272	10 i.U. Lente
	15. 3. 68	270	22 i.U. Lente

Total IRI in insulin-treated patients. Table 2
shows the results of total-IRI determinations in sera
from 169 insulin-treated fasting diabetic patients. The
serum was drawn 14—24 h after the last insulin injection. All patients had been treated with insulin for

more than six months. The majority, by far, had elevated total IRI levels as compared to normals and untreated patients. Two examples of the variation in fasting total IRI in the course of treatment are shown in Table 3. A marked increase in total IRI was observed after 2—3 months of treatment.

Table 4. *Variation in total fasting IRI in diabetics who had been treated with insulin for several years*

Patient	Treatment started	Date	Total IRI	Daily insulin dose i. U. Lente
A.B.J.	1951	1. 9. 66	2625	44
		28. 4. 67	1915	50
		24. 8. 67	2550	46
		24. 11. 67	1540	46
		2. 2. 68	1452	46
		3. 5. 68	1500	—
H.L.T.	1945	18. 11. 66	1128	48
		25. 11. 66	1074	44
		2. 12. 66	1120	42
J.Å.H.	1936	7. 10. 66	224	32
		14. 10. 66	190	34
		21. 10. 66	233	34
Å.H.	1942	26. 9. 66	346	28
		7. 10. 66	214	28
H.S.	1951	26. 9. 66	2212	48
		15. 10. 66	1700	48
E.W.	1964	24. 9. 66	23	32
		26. 9. 66	43	32
J.H.	1937	7. 10. 66	280	36
		14. 10. 66	380	36
		31. 10. 66	352	36
B.W.	1940	7. 10. 66	318	30
		14. 10. 66	312	30
F.M.	1965	6. 10. 66	54	24
		14. 10. 66	94	24
		21. 10. 66	61	24
F.J.[a]	1950	5. 5. 67	35150	184
		16. 5. 67	50700	240

[a] serum from this resistant patient was obtained by courtesy of Dr. Hockaday, Oxford, England.

The variation in total IRI in patients who had been treated with insulin for several years is shown in Table 4. The level of total fasting IRI remained fairly stable from week to week.

The daily insulin injection(s) caused some variation in the level of total IRI, as shown in Fig. 6.

Discussion

1. *Insulin radioimmunoassay*. Each radioimmunoassay method has a number of sources of error, some of which are common to most methods, others being specific for a particular method.

The adsorption of ^{125}I-insulin to plastic and glassware is probably of much greater consequence than one usually realizes. It has been generally accepted that the presence of 0.1% albumin effectively prevents this adsorption of ^{125}I-insulin. However, this is not the case in every type of tube. The ability of ^{125}I-insulin

to concentrate in foam may completely invalidate the results.

Variation in the protein concentration will influence most of the separation techniques used in radioimmunoassay. Thus, the binding of free ^{125}I-insulin to cellulose, ion-exchangers, charcoal and similar substances is diminished at higher albumin concentrations. When ethanol is used for the separation of free and bound ^{125}I-insulin, the coprecipitation of free ^{125}I-insulin will increase at higher protein concentrations. It is therefore obvious that the protein concentration in the standards must be the same as in the test samples.

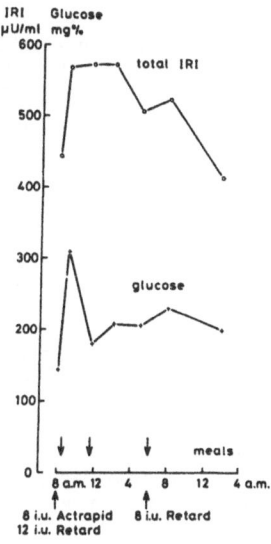

Fig. 6

Enzymatic degradation of insulin during incubation will, of course, give false values. The degree of degradation in human sera was found to be very low, but sera from other species exhibited a much more pronounced degradation of ^{125}I-insulin. Some types of albumin contain enzymes that degrade the standard insulin (Fig. 4) and thereby reduce the stability of the standard solutions. When use was made of a pure albumin, solutions containing from 10 to 5000 μU insulin/ml were found to remain stable for more than a year at −18°C. Anti-insulin sera stored at −18°C have been used since 1966 without our detecting any changes in their binding capacity.

The immunoassay method here described makes use of 96% ethanol for the separation of free and antibody-bound ^{125}I-insulin. This method has several advantages: it is accurate, quick and easy to perform, and the 96% ethanol is an ordinary, inexpensive standard chemical. For optimal results a good ^{125}I-insulin preparation is required, as shown in Table 1. Ethanol separation has yielded results identical with those of the chromatographic and double-antibody separation techniques.

2. Determination of total IRI in serum from insulin-treated persons. The presence of antibodies and anti-body-bound insulin in serum from insulin-treated persons necessitates carrying out an acid ethanol extraction to obtain an estimate of the total IRI. The method involves a complete dissociation of the insulin-antibody complex, followed by a precipitation of the antibodies in ethanol at neutral pH. This method offers several advantages as compared to separation by gel filtration described by Pearson and Martin (1970) — it is a routine method in which more than 50 samples can easily be handled simultaneously; the dry residue can be dissolved direct in the buffer used for immunoassay.

The results obtained with this method showed the following:

1. Total fasting IRI in a randomly selected diabetic population treated with insulin in doses of between 6 and 120 units/day varied from 6 to 4347 $\mu U/ml$ (mean 392 $\mu U/ml$), and by far the majority of the patients had substantially elevated IRI values when compared with normal subjects and diabetics not treated with insulin.

2. The total IRI increased during the period of treatment from the normal values of the first month or two to a higher level, which then became fairly stable after about 5 months of treatment (two examples are shown in Table 3). The increase in IRI was observed simultaneously with the formation of antibodies. If no antibodies were produced, for instance due to treatment with monocomponent-insulin (Schlichtkrull *et al.*, 1971), the serum IRI stayed within normal limits.

3. The fasting total IRI in diabetic patients treated with insulin for over five months varied only a little from week to week. Even in those treated several months longer, the changes observed were minor (Tables 3 and 4). That is to say, each patient has his own total IRI level.

4. Insulin-resistant patients (defined as diabetics whose daily dose of insulin exceeds 100 units) had so far proved to have extraordinarily high levels of IRI (one such case is demonstrated in Table 4).

5. The daily insulin injection(s) caused variations in the IRI level (Fig. 6).

The composition of the total IRI is of great interest. Most of the IRI is, naturally, bound to antibodies and has hardly any biological effect *in vivo* or *in vitro*. Stout and Vallance-Owen (1969) put forward the hypothesis that hyperinsulinism plays a major role in the development of vascular disease. Stout (1970) showed that chickens treated with insulin for 19 weeks developed vascular lesions faster than untreated chickens. Thus, the high total serum IRI found in diabetics could be contributory to the development of vascular complications in diabetes.

Acknowledgements. I wish to thank Jørgen Schlicht-krull, D. Sc., for his interest, constructive criticism and valuable advice; Mrs. Ulla Dahl Larsen for preparing the excellent ^{125}I-insulin; the staff of Hvidøre Diabetes Hospital for drawing most of the serum samples used in this study; Mrs. Majken Petri Petersen, Mrs. Connie Eriksen and Miss Lene Hansen for their excellent technical assistance; and Mrs. Birgit Jensen for drawing the curves.

References

Berson, S.A., Yalow, R.S.: The present status of insulin antagonists in plasma. Diabetes **13**, 247—259 (1964).
Grodsky, G.M.: Production of auto-antibodies to insulin in man and rabbits. Diabetes **14**, 396—403 (1965).
Heding, L.G.: A simplified insulin radioimmunoassay method, in "Labelled Proteins in Tracer Studies", Ed. L. Donato et al. Euratom, Brussels 345—350 (1966).
— Nielsen, A. Vølund: Determination of free and anti-body-bound IRI in serum from insulin-treated diabetic patients. Abst. in Exerpta Medica international congress series No. 140, Sixth Congress of the International Diabetes Federation, Stockholm, 1967.
— Determination of free and antibody-bound insulin in insulin-treated diabetic patients. Horm. Metab. Res. **1**, 145—146 (1969).
Jørgensen, K., Binder, C.: ^{125}I-insulin as a tracer of insulin in different chemical processes, in "Labelled Proteins in Tracer Studies", Ed. L. Donato et al. Euratom, Brussels, 329—333 (1966).
Kirkham, K.E., Hunter, W.M.: Radioimmunoassay Methods, Churchill Livingstone, Edinburgh and London (1971).
Ohneda, A., Toyota, T., Sato, S., Yamagata, S.: Extraction of plasma insulin in radioimmunoassay for removal of nonspecific inhibitor and of circulating insulin antibody. Tôhoku J. exp. Med. **100**, 75—84 (1970).
Pearson, M.J., Martin, F.I.R.: The separation of total plasma insulin from binding proteins using gel filtration: its application to the measurement of rate of insulin disappearance. Diabetologia **6**, 581—585 (1970).
Schlichtkrull, J., Heding, L.G., Christiansen, Aa.H., Vølund, Aa.: Immunological aspects of insulin therapy. Paper presented at the VIIth Acta endocrinologica Congress, Copenhagen, Denmark (1971).
Stout, R.W., Vallance-Owen, J.: Insulin and atheroma. Lancet **1969** I, 1078—1080.
— Development of vascular lesions in insulin-treated animals fed a normal diet. Brit. med. J. **1970 III**, 685—687.
Vølund, A.: Computerized calculation and control of radioimmunoassay or bioassays illustrated for the insulin assay. Paper to be published (1972).
Yalow, R.S., Berson, S.A.: Immunoassay of endogenous plasma insulin in man. J. clin. Invest. **39**, 1157—1175 (1960).

Dr. L.G. Heding
Novo Research Institute
115, Fugglebakkevej
2200 Copenhagen-N
Denmark

42

Horm Metab Res (Suppl) 1974; 5:40–4.

Radioimmunoassays for Human, Pork and Ox C-peptides and Related Substances

L.G. Heding, U.D. Larsen, J. Markussen, K.H. Jørgensen and O. Hallund

Novo Research Institute, Novo Alle, DK-2880 Bagsværd.

Summary

Antibodies reacting with ox, pork and human C-peptide have been prepared using ox and pork proinsulin and human b-component as the immunogens. Specific assays for C-peptide and/or proinsulin-like substances were established by using ^{125}I-Tyr-C-peptide as tracer. In solutions (serum and pancreatic extracts) containing C-peptide, proinsulin and insulin, the two latter substances were bound by addition of insulin antibodies and separated from C-peptide by ethanol precipitation. Thus the determination of C-peptide could be performed on the supernatant without interference from proinsulin. Pancreatic extracts were found to contain equimolar amounts of insulin and C-peptide, whereas proinsulin constituted about 3% (molar) of the total IRI. After i.v. injections of glucagon, equimolar amounts of insulin and C-peptide were found to be secreted into venae pancreatico-duodenales. The average concentration of C-peptide in serum from 9 normal persons was found to be 0.33 ± 0.05 pmole/ml. After oral administration of 1.75 g glucose per kg, the value increased to 2.3 ± 0.3 pmole/ml.

Key words: Insulin – Proinsulin – C-peptide – Radioimmunoassay-Porcine, Bovine, Human.

Introduction

Since the discovery of proinsulin by *Steiner and Cunningham* (1967) and the subsequent isolation of proinsulin, intermediate forms and other minor components present in crystalline insulin (*Chance, Ellis and Bromer* 1968; *Steiner, Hallund, Rubenstein, Cho and Bayliss* 1968), a series of papers have been published concerning proinsulin immunoassays (*Yip and Logothetopoulos* 1969; *Rubenstein, Steiner, Cho, Lawrence and Kirsteins* 1969; *Stoll, Touber, Ensinck and Williams* 1970; *Melani, Rubenstein, Oyer and Steiner* 1970).

The differences between C-peptides, and consequently between proinsulins of various species, are so extensive that each species requires its own immunoassay. Furthermore, for each species, specific determination of the three substances, insulin, C-peptide and proinsulin-like substances (in the following referred to as "proinsulin") in samples containing them all (e.g., serum or pancreas) requires: (I) separation of C-peptide from insulin and "proinsulin", (II) two antisera - a serum that reacts with the insulin sites and a serum that reacts with the C-peptide sites, and (III) labelled insulin and C-peptide as tracers. The separation step (I) can be left out when samples contain only two of the substances.

Several investigators have used gel filtration for the separation of insulin and proinsulin (*Melani* et al. 1970; *Steiner* et al. 1968; *Schmidt* 1969) others employed polyacrylamide gel electrophoresis (*Gutman, Lazarus and Recant* 1972). All these methods are time-consuming because they yield a number of fractions from each sample. This paper describes (I) a simple and quick method for separation of C-peptide from insulin and "proinsulin", (II) preparation of human, porcine and bovine proinsulin antisera raised in guinea pigs and (III) preparation of ^{125}I-Tyr-C-peptide. By these means, it was possible to determine specifically C-peptide in pancreatic extracts, C-peptide and "proinsulin" in human, porcine and bovine serum, and proinsulin immunoreactive substances in various insulin preparations in the presence of more than 100,000 times as much insulin.

Materials and Methods

Bovine and porcine proinsulin were prepared from b-component by anion exchange chromatography. About 6 g of b-component was applicated on a column (5 x 53 cm) of QAE-Sephadex A-25 equilibrated with a buffer (0.1 M NH$_4$Cl; 0.02 M NH$_3$ in 60% (v/v) ethanol, pH 8.4 at 25°C) and eluted with the same buffer at a rate of 120 ml/h and at 25°C. Fractions of 15 ml were collected. Proinsulin was isolated from the second main peak eluting at approximately 2 column volumes, by evaporation, salting out, desalting on Sephadex G 25 and freeze-drying. The purity was checked by polyacrylamide disc gel electrophoresis and isoelectric focusing and found to be about 97%. Bovine and porcine intermediate forms were obtained from *J. Brange*, the Novo Insulin Laboratory. The purity of these substances was checked as described for proinsulin, and found to be approximately 95%. Human b-component was obtained by gelfiltration of first crystals of human insulin on Sephadex G-50. 140 ng of purified human proinsulin was obtained from Dr. D.F.Steiner, Chicago. Bovine and porcine C-peptides were isolated and purified according to *Sundby and Markussen* (1970). Human C-peptide was prepared according to *Markussen, Sundby, Smyth and Ko* (1971) with the following modifications: The product obtained after the fourth column was further purified on an anion exchange (AG 1 x 2, Bio Rad) column (1.2 x 44 cm) equilibrated with 0.1 M pyridinium acetate adjusted to pH 5.0 with pyridine. The column was eluted using a pH-gradient from 5.0 to 3.7 produced by 250 ml pH 5 buffer, and 250 ml buffer adjusted to pH 3.7 with acetic acid. The chymotryptic fragment, residues 1-24 of human C-peptide eluted at pH 3.8, while intact C-peptide could first be eluted by changing the eluent to 1 M formic acid. The materials were isolated by freezedrying. Synthetic human C-peptide for standard solutions was obtained from Dr. V.K.Naithani*,Aachen. Tyrosylated pork and ox C-peptides were prepared as described by *Markussen, Jørgensen and Heding* (1970). Human Tyr-C-peptide was prepared as follows: 60 mg of tBOC Tyr-p-nitrophenyl ester in 5 ml DMF (dimethyl formamide) was mixed with 8 mg C-peptide in 2 ml H_2O. pH was adjusted to and maintained at 8.6 with 1 M N-methyl morpholine in DMF. After 72 h of incubation at 4°C, excess N-methyl morpholine was evaporated and pH adjusted to 2.5 with acetic acid (conc.). The solution was applied to a SP-Sephadex C-25 column, equilibrated with 0.5 M acetic acid, which was also used as eluent. tBOC-Tyr-C-peptide eluted in the void volume, while unreacted C-peptide was retarded by ion exchange and tBOC-Tyr was retarded by gelfiltration. After freezedrying of the tBOC-Tyr-C-peptide pool, the tBOC group was split off by treatment with 0.5 ml 3.5 M HCl in dioxane for 0.5 h. After evaporation, the precipitate was dissolved in 100 μl H_2O and precipitated by addition of 6 ml acetone. Finally, the Tyr-C-peptide was isolated by centrifugation and dissolved in 2 ml H_2O. The Tyr-C-peptides were iodinated with ^{125}I, using a modification of the method described by *Markussen* et al. (1970). 100 μl of a solution containing 20 μg of Tyr-C-peptide was pipetted into a small test tube and mixed with 10 μl of 1 N HCl and 10 μl containing in the order of 1 mCi of a carrier-free solution of $^{125}I^-$ (Amersham). 25 μl of a 2% (w/v) solution of KIO_3 was added while stirring, and the iodination was stopped after 10 minutes by addition of 150 μl of 0.1 M $NaHSO_3$. Furthermore, 500 μl of distilled water and 200 μl of 0.1 M histidine were added to adjust the pH to 6.1. The solution was then subjected to anion exchange chromatography on a column (d = 0.7 cm) containing 50 mg of QAE-Sephadex A 25 equilibrated with a 0.1 M histidine, 0.05 N HCl buffer, pH 6.1. To remove unreacted $^{125}I^-$, the column was eluted with the latter histidine buffer (at a rate of about 0.5 ml/min.), and 1 ml fractions were collected until the radioactivity reached and remained at a low level (after about 20 fractions). The ^{125}I-Tyr-C-peptide was then eluted with 1% human albumin (Behringswerke, trocken, reinst) in 0.02 N HCl. The solution was diluted with 1% human albumin to make 20 ml, corresponding to a calculated concentration of approximately 1 μg of labelled and unlabelled Tyr-C-peptide per ml. Finally, the pH was adjusted to about neutrality with 1 N NaOH. The over-all radioactivity yield varied between 20 and 56% and the calculated specific radioactivity between 7 and 26 mCi/mg.

Antibodies against ox and pork proinsulin were raised in guinea pigs injected subcutaneously once every third week with 2 ml of an emulsion of equal volumes of Freund's adjuvant and of a proinsulin

---

* Deutsches Wollforschungsinstitut

solution containing 40 μg/ml in 0.04 M phosphate buffer, pH 7.4. Complete Freund's adjuvant was used for the first injection, in complete adjuvant for the subsequent injections. Antibodies against human C-peptide were raised by s.c. injection once every third week with 0.5 ml of an emulsion of equal amounts of Freund's adjuvant and of b-component solution containing approx. 2 mg/ml. Five to eight ml blood samples were drawn by cardial puncture 14 days after the third injection and then at regular monthly intervals.

Insulin was determined by immunoassay as described by *Heding* (1972), using ethanol for the precipitation of the antigen-antibody complex. The proinsulin immunoassay was a modification of this method: ^{125}I-Tyr-C-peptides and antisera were diluted in phosphate buffer (0.04 M, pH 7.4) containing human albumin (1 g/l) and thiomersal (0.2 g/l), in the following referred to as FAM. All standards and samples were diluted in a phosphate buffer (0.04 M, pH 7.4) containing NaCl (6 g/l), human albumin (10 g/l) and thiomersal (0.2 g/l), in the following referred to as NaFAM. The radioimmunoassay was performed as follows: to triplicates of 100 μl of standard solutions (containing, e.g., 1-10 ng of proinsulin, intermediate forms or C-peptide per ml) or samples was added 100 μl of diluted anti-proinsulin serum. After 20 h of incubation at 4°C, 100 μl of ^{125}I-Tyr-C-peptide (1-2 ng/ml) was added at 4°C, and after a further 24 h of incubation at 4°C, 1.6 ml of 96% ethanol was added at 4°C. After mixing on Vortex and centrifugation, the supernatant was discarded and the precipitate was washed once with 2 ml of 80% ethanol-buffer (18 ml FAM + 162 ml of distilled water + 960 ml of 96% (v/v) ethanol). Finally, the precipitate was dissolved in 0.6 ml of 0.05 N NaOH and counted.

Separation of C-peptide from insulin plus insulin-like substances (such as proinsulin) in serum or pancreatic extracts was performed as follows: To 0.5 ml samples was added 0.1 ml of anti-insulin guinea pig serum (1:10) corresponding to a binding capacity of 30 mU. After one hour of incubation, 2.5 ml of ethanol was added, and after shaking and centrifugation, the ethanolic supernatant containing the C-peptide was decanted from the precipitate which then contained insulin and insulin-like substances bound to antibodies.

The supernatant was transferred to a small bottle and evaporated to dryness in a dessicator at 10 mm Hg overnight. The residue was dissolved in NaFAM and immunoassayed for C-peptide. When "proinsulin" was to be determined, the following procedure was used: To 1 ml of serum or diluted extract was added 1.8 ml of ethanol. After shaking and centrifugation, the supernatant containing the C-peptide, insulin and "proinsulin" was evaporated to dryness and dissolved in 0.5 ml of buffer. Incubation with anti-insulin serum and precipitation with ethanol was then performed as described above. From the precipitate, insulin and insulin-like substances were extracted as described by *Heding* (1972), and proinsulin determined

44

in the C-peptide immunoassay. Recovery of added pork proinsulin was found to be approx. 60%.

Normal humans were given 1.75 g glucose per kg ideal body weight after an overnight fast, and blood samples were taken before and one hour after glucose. One pig (100 kg) was anesthetized and a catheter placed in vena pancreatico-duodenalis. 0.1 mg glucagon/kg was injected i.v. four times at one hour intervals, and blood obtained at 0, 2, 5, 10, 20, 40 and 60 minutes after each injection. Furthermore, peripheral vein blood was obtained from a cow and a pig.

Results

Binding of ^{125}I-Tyr-C-peptide to antibodies.

All guinea pigs injected with 40 μg ox or pork proinsulin produced antibodies. The serum could be used for immunoassay in dilutions 1:1500 - 1: 6000. When 1 mg amounts were injected, the sera could be used in dilutions 1:50.000. Guinea pigs immunized with the human b-component developed antibodies reacting with human ^{125}I-Tyr-C-peptide. All groups of guinea pigs developed antibodies against insulin. An excess of proinsulin antibodies was capable of binding between 66 and 93% of the radioactivity in the ^{125}I-Tyr-C-peptides of pork and ox, and only 62% of the human ^{125}I-Tyr-C-peptide. With no antibodies present, only 1-3 percent of the labelled C-peptides were precipitated by ethanol.

Displacement of antibody-bound ^{125}I-Tyr-C-peptide by proinsulin, intermediate forms and C-peptide.

Figures 1, 2 and 3 show the standard curves in the human, ox and pork systems respectively. Specially purified insulin was found incapable of displacing measurable amounts of labelled ox and pork C-peptide even in concentrations as high as 100.000 ng/ml.

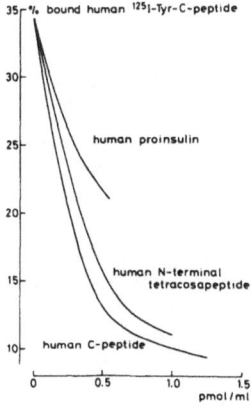

Fig. 1 Standard curves with human synthetic C-peptide and proinsulin.

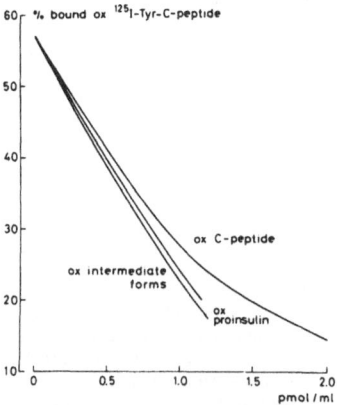

Fig. 2 Standard curves with bovine C-peptide, proinsulin and intermediate forms.

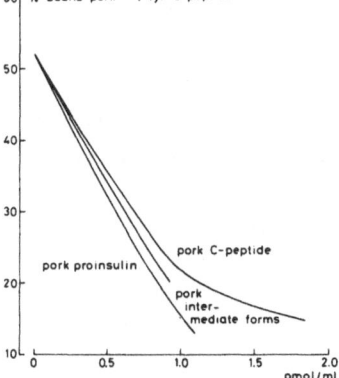

Fig. 3 Standard curves with porcine C-peptide, proinsulin and intermediate forms.

Proinsulin and the intermediate forms reacted equally well in the bovine and porcine systems (Fig. 2, 3).

The chymotryptic fragment, (1-24), of human C-peptide reacted approximately 20% less than the synthetic C-peptide.

Separation of proinsulin + insulin from C-peptide in pancreatic extracts.

Before estimating the C-peptide in the supernatant after removal of insulin plus proinsulin by insulin antibodies, a check was made to ensure that all insulin immunoreactivity had been removed. No residual IRI was ever found in the supernatant. The recovery of added C-peptide was between 90 and 100%.

Determination of C-peptide and proinsulin in pancreatic extracts.

In acid ethanol extracts of pancreas (in the beginning

Table 1 C-peptide and IRI in serum

| | pmole/ml | | | |
| | Fasting | | 1 h after glucose | |
	C-peptide	IRI	C-peptide	IRI
9 N-persons (mean ± 1 SEM)	0.33 ± 0.05	0.085 ± 0.022	2.3 ± 0.3	0.53 ± 0.06
1 pig	0.67	0.13	–	–
1 cow	0.50	0.07	–	–

of insulin manufacture), insulin and C-peptide were found in nearly equimolar amounts, e.g., 29 μmole of insulin and 32 μmole of C-peptide per kg of ox pancreas. The amount of proinsulin and proinsulin-like immunoreactive substances in the pancreas was approximately 3% of the insulin content on a molar basis. In the mother liquid from the salting-out process, the concentration of C-peptide was found to be much higher than that of IRI, e.g., 0.07 nmole of IRI and 7 nmole of C-peptide per ml. This solution has been used as a starting material in the preparation of C-peptide (*Sundby and Markussen* 1970).

C-peptide in serum

Table 1 shows that C-peptide is present in fasting serum. In all individuals, the molar ratio between C-peptide and insulin was found much higher than 1 in spite of the equimolar amounts found in the pancreas. After stimulation by glucose, both insulin and C-peptide increase.

Figure 4 shows the results of repeated i.v. glucagon injections into a pig on the secretion of insulin and

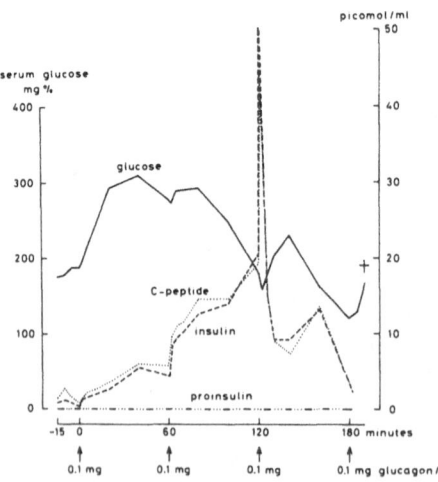

Fig. 4 Repeated i.v. glucagon injections (0.1 mg/kg) into a pig, blood sampling from vena pancreatico duodenale.

C-peptide into venae pancreatico-duodenales. It appears that the two substances are secreted in equimolar amounts through the whole period. Furthermore, only 0.01-03 pmole/ml of "proinsulin" was detected, meaning that less than 3% of the IRI was proinsulin on a molar basis.

Discussion and Conclusion

In establishing a specific radioimmunoassay for C-peptide and/or proinsulin, the use of C-peptides rather than proinsulin for the production of antibodies seems logical to eliminate the simultaneous production of insulin antibodies. However, C-peptide has proved to have low immunogenicity even when coupled to proteins (*Markussen, Heding, Jørgensen and Sundby* 1971), and only a few papers report on the production of antibodies towards C-peptide (*Melani* et al. 1970, and *Kaneko, Oka, Oda, Yanaihara, Hashimoto and Yanaihara*, 1973). The present report demonstrates that antibodies against C-peptide can be prepared using ox and pork proinsulin. It was also possible to raise antibodies against human C-peptide using human b-component as immunogen. As little as 40 μg injected every third week is sufficient to raise antibodies, but higher doses increase the capacity. The presence of insulin antibodies does not interfere with the C-peptide assay when ^{125}I-Tyr-C-peptide is used, since only C-peptide and proinsulin-related substances can compete with this tracer.

In order to determine C-peptide and not the sum of C-peptide + "proinsulin", the removal of insulin + "proinsulin" from serum with a surplus of insulin antibodies followed by ethanol precipitation was found quick and reliable and left virtually all C-peptide in the supernatant. This method greatly increased the number of samples which could be handled.

In the pancreas equimolar amounts of C-peptide and insulin and approximately 3 mole-% proinsulin was found. After i.v. glucagon injections into a pig, insulin and C-peptide was secreted in equimolar amounts in venae pancreatico-duodenales, while "proinsulin" constituted less than 3% of the total IRI.

In peripheral blood, the molar concentration of C-peptide was found higher than that of insulin.

This contrasts with the original finding of *Melani* et al., 1970, reporting equimolar amounts, but is in accordance with a later paper (*Block, Nako, Steiner* and *Rubenstein* 1972), where higher values for C-peptide were reported. As C-peptide has a lower metabolic clearance rate than insulin (*Katz and Rubenstein* 1973), a molar ratio higher than one between C-peptide and insulin is to be expected.

Aknowledgement

We wish to tank Dr. J. Schlichtkrull for valuable suggestions and criticism, Mr. J. Brange of the Novo Insulin Laboratory for the intermediates, Mr. A.S. Fries for performing the blood sampling from the pigs, Dr. D.F. Steiner, Chicago, for the human proinsulin, Dr. V.K. Naithani, Aachen, for the synthetic human C-peptide and Mrs. C. Eriksen for performing the major part of the radioimmunoassays.

References

Block, Marshall B., M.E. Mako, D,F. Steiner, A. Rubenstein: Circulating C-peptide immunoreactivity. Studies in normals and diabetic patients. Diabetes 21: 1013-1026, 1972

Chance, R.E., R.M. Ellis, W.W. Bromer: Porcine proinsulin: Characterization and amino acid sequence. Science 161: 165-167, 1968

Gutman, R.A., N.R. Lazarus, L. Recant: Electrophoretic characterization of circulating human proinsulin and insulin. Diabetologia 8: 136-140, 1972

Heding, L.G.: Determination of total serum insulin (IRI) in insulin-treated diabetic patients. Diabetologia 8: 260-266, 1972

Kaneko, T., H. Oka, T. Oda. N. Yanaihara, T. Hashimoto, C. Yanaihara: Human plasma C-peptide immunoreactivity in the healthy and diabetes. Abstract N°39, Excerpta Medica, N° 280, VIII Congress of the International Diabetes Federation, Brussels Belgium, July 15-20, 1973

Katz, A.I., H. Rubenstein: Metabolism of proinsulin, insulin and C-peptide in the rat. J. clin. Invest. 52: 1113-1121, 1973

Markussen, J.. K.H. Jørgensen, L.G. Heding: Preparation of bovine ¹²⁵I-Tyrosyl-C-peptide. Horm. Metab. Res. 2: 53-55, 1970

Markussen, J., L.G. Heding, K.H. Jørgensen, F. Sundby: Proinsulin, insulin and C-peptide. Horm. Metab. Res., Suppl. series N° 3, Symposium in Ulm, February 25, 1970. Thieme, Stuttgart 1971

Markussen, J., F. Sundby, D.G. Smyth, A. Ko: Preparation of human C-peptide. Horm. Metab. Res. 3: 229-232, 1971

Melani, F., A.H. Rubenstein, Ph. E. Oyer, D.F. Steiner: Identification of proinsulin and C-peptide in human serum by a specific immunoassay. Proc. Nat. Acad. Sci. 67: N° 1, 148-155, 1970

Rubenstein, A.H., D.F. Steiner, S. Cho, A.M. Lawrence, L. Kirsteins: Immunological properties of bovine proinsulin and related fractions. Diabetes 18: 598-605, 1969

Schmidt, D.: Proinsulin. Postgrad. med. J. 45: 482-487, 1969

Steiner, D.F., D. Cunningham: Insulin biosynthesis: Evidence for a precursor. Science 157: 697-700, 1967

Steiner, D.F., O. Hallund, A. Rubenstein, S. Cho, C. Bayliss: Isolation and properties of proinsulin, intermediate forms, and other minor components from crystalline bovine insulin. Diabetes 17: N° 12: 725-736, 1968

Stoll, R.W., J.L. Touber, J.W. Ensinck, R.H. Williams: Substances immunologically related to proinsulin or conecting peptide in swine plasma. Horm. Metab. Res. 2: 153-156, 1970

Sundby, F., J. Markussen: Preparative method for the isolation of C-peptides from ox and pork pancreas. Horm. Metab. Res. 2: 17-20, 1970

Yip, C.C., Logothetopoulos, J.: A specific anti-proinsulin serum and the presence of proinsulin in calf serum. Proc. Nat. Acad. Sci. 65: 415-419, 1969

Diabetologia 11, 541–548 (1975)
© by Springer-Verlag 1975

Radioimmunological Determination of Human C-Peptide in Serum

L. G. Heding

Novo Research Institute, Copenhagen, Denmark

Received: February 20, 1975, and in revised form: August 18, 1975

Summary. A routine radioimmunoassay for human C-peptide in serum is described. Antibodies against human C-peptide were raised by immunizing guinea pigs with human b-component. Nine out of 12 animals produced useful antibodies within 6 months. Insulin antibodies coupled to Sepharose were used to bind human proinsulin and insulin in the serum and after centrifugation C-peptide was determined in the supernatant. The detection limit of the assay (calculated as 2 SD from zero) was about 0.003 pmole of C-peptide (in 100 μl). The main sources of error were: (1) Normal and diabetic sera devoid of C-peptide gave a displacement of ^{125}I-Tyr-C-peptide varying from 0 to 0.16 nM (6 different antisera). Only one antiserum (M 1181) showed no displacement, and the values of C-peptide determined with this antiserum in normal and diabetic sera were lower than the values determined with another antiserum, which gave a value of 0.07 nM in the sera free of C-peptide. It is suggested

that displacement found with most antisera is due to substances in serum that are not related to C-peptide or proinsulin. (2) Serial dilutions of pancreatic extracts and sera may yield dilution curves slightly different to those of the synthetic standard. Possible explanations are discussed. These sources of error can be eliminated or reduced by the proper selection of antisera. Fasting sera from 15 normals, 8 maturity-onset diabetics and 10 insulin-requiring diabetics showed the following concentrations of C-peptide: (M 1181) 0.35 ± 0.09, 0.74 ± 0.51 and 0.21 ± 0.14 (nM, mean ± SD). One hour after 1.75 g/kg oral glucose the values increased to 2.24 ± 0.71, 2.34 ± 1.18 and 0.42 ± 0.24 nM.

Key words: Radioimmunoassay, human C-peptide, human proinsulin, insulin.

A sensitive and specific immunoassay for routine determination of human C-peptide in serum requires a high-affinity antibody and a simple and quick method for the separation of C-peptide from proinsulin. Unfortunately, C-peptides (molecular weight about 3000) show low immunogenicity [10] in spite of the great difference in the amino acid sequences of the injected material and of the endogenous C-peptide of the immunized animal. An assay of human C-peptide in serum has been described [14] that comprised extraction of serum, gel filtration of the serum extract and analysis of the individual fractions after evaporation, using an antiserum raised against the natural human C-peptide coupled to albumin. Recently, total C-peptide (CPR) immunoreactivity (C-peptide + proinsulin + intermediates) were determined in unextracted serum [2]. This method, in contrast to [14], yielded higher molar concentrations of C-peptide than those of insulin. This was in agreement with the findings in ox serum [16] and in human sera [6]. An assay based upon antibodies against synthetic human connecting peptide has also been described [8]. As a consequence of its low immunogenicity, antibodies to human C-peptide show low affinity, which implies a high detection limit, and the best result yet reported was the recording of 0.05 ng per tube, corresponding to about 0.25 ng/ml sample [8]. A method has been reported [6] of raising antibodies of high affinity to

human C-peptide, using human b-component for immunization and a quick method of separation of C-peptide from proinsulin + intermediates (henceforth referred to as PLI). The present paper describes a new and simplified method of separation of proinsulin + insulin + intermediates from C-peptide, using antibodies against pork insulin coupled to Sepharose. The advantage of this separation over the separation techniques formerly described is that, besides being easier and simpler, this technique enables one to isolate proinsulin from, e. g., 2 ml of serum.

Materials and Methods

Preparation and Testing of Antibodies to Human C-Peptide

Human C-peptide was prepared according to [9] and tyrosylated and iodinated as described in [6]. Twelve guinea pigs were immunized with 1 ml of an emulsion consisting of equal volumes of Freund's adjuvant (complete in the first injection, incomplete in subsequent injections) and a solution of crude human b-component (2 mg/ml) obtained by gel filtration of first crystals of human insulin on Sephadex G 50. The animals were injected every third week. Blood was drawn by cardiac puncture, initially 14 days after the fourth injection, then regularly 14 days after each subsequent injection. Human ^{125}I-Tyr-C-peptide and antisera were diluted in phosphate buffer (0.04 M, pH 7.4) containing human albumin (Behringwerke, electrophoretic purity 100%) (1 g/l) and thiomersal (0.2 g/l), henceforth referred to as FAM. All standards and samples were diluted in phosphate buffer (0.04 M, pH 8.4) containing NaCl (6 g/l), human albumin (60 g/l) and thiomersal (0.2 g/l) (NaFAM). The antisera were tested in the following manner: to duplicated of 100 μl of diluted serum were added 100 μl of ^{125}I-Tyr-C-peptide (2 ng/ml) and 100 μl of Na-FAM. After 20–24 hrs of incubation at 4° C, free and antibody-bound ^{125}I-Tyr-C-peptide was separated by addition of 1.6 ml 95% (v/v) ethanol at 4 ° C. After centrifugation and one wash the precipitate was dissolved in 0.6 ml of 0.05 N NaOH and counted as described for glucagon [5].

Preparation and Testing of Insulin Coupled to Sepharose (S-I)

Monocomponent (MC) insulin (ox or pork) was coupled to Sepharose as follows: 50 g of Sepharose 4 B slurry was suspended in 10 ml of distilled water. 10 g of CNBr was added to the suspension and the pH maintained at 10.5 for 15 min with 4 N NaOH. The activated Sepharose was filtered off and washed with 0.5 l of 0.2 M NaHCO$_3$, and then sucked dry. The CNBr-activated Sepharose 4 B was then suspended in a solution of MC insulin prepared as follows: 2 g of insulin was dissolved in 100 ml of distilled water acidified by addition of 1 N HCl to a pH of approximately 2.5. 100 mg of Na$_2$EDTA was added and the pH raised to approximately 7 with 1 N NaOH. The insulin solution was mixed with 100 ml of 0.2 M NaHCO$_3$, pH 8.0, and the pH adjusted to 8.0. The reaction mixture of activated Sepharose and insulin solution was stirred at 4° C overnight. Absorbance at 278 and 250 nm was measured before and 1, 3 and 20 h after the start of reaction. About 20% of the insulin was found to be bound to the Sepharose after 1 h, whereas no further increase in binding was found between 1 and 20 h. After binding of the insulin to Sepharose, 50 ml of 5 M redistilled ethanolamine adjusted to pH 8.1 with conc. HCl was added to the suspension in order to inactivate the residual reactive groups on the activated Sepharose. After 64 h of stirring at 4° C, the Sepharose was filtered off and washed with 0.5 l each of the following solutions: distilled H$_2$O, 0.5 M NaCl in 60% ethanol, and NaFAM. Finally, the Sepharose with insulin coupled to it (in the following referred to as S-I) was sucked dry and suspended in NaFAM, pH 7.4, to a concentration of 1 mg bound insulin per ml.

The immunoreactive insulin (IRI) of the S-I was determined as follows: To duplicated of 100 μl of S-I suspension were added increasing amounts of guinea pig antiinsulin serum with known binding capacity. After 20 h at 4°C with shaking, the S-I was removed by centrifugation and the insulin binding capacity determined in the supernatant, using ^{125}I-pork insulin. The insulin antibodies bound by the S-I were calculated and referred to as the IRI of the S-I.

Preparation of Antibodies to Insulin, Coupling to Sepharose (S-AIS) and Binding Capacity of S-AIS

Antibodies to insulin were isolated as follows: guinea pig antipork insulin serum was prepared by repeated immunization as described in [4]. Guinea pig serum with a known insulin-binding capacity (in the range of 4 to 20 units) was mixed with a suspension of S-I containing a surplus of IRI (8–40 units). The mixture was shaken overnight at 4° C, filtered in a small glass tube containing glass paper, and washed three times with 1 ml 0.6% NaCl. The insulin antibodies were eluted from the S-I column with 2 ml portions of dilute HCl (0.005–0.001 N) at 0° C. The 2 ml eluates were run direct into tubes containing 2 ml 0.04 M phosphate buffer (NaFAM) to reduce the time of exposure of the antibodies to a low pH. The antibody-containing eluates were pooled, freeze-dried and stored at −18° C until used.

The purified insulin antibodies were coupled to CNBr-activated Sepharose 4 B in the following manner: the freeze-dried antibodies were dissolved in 25 ml of distilled water to a concentration of approximately 0.5 unit/ml (calculated: 0.48 U/ml; checked by analysis: 0.51 U/ml), 400 mg of NaHCO$_3$ and 200 mg of NaCl were added and the pH adjusted to 8.0. The solution was turbid. A quantity of freshly prepared CNBr-activated Sepharose 4 B corresponding to 24 g Sepharose was suspended in the antibody solution and the reaction mixture allowed to stand at 4° C overnight with stirring. 24 ml of redistilled ethanolamine adjusted to a pH of 8.0 with conc. HCl was added, the mixture stirred at 4° C for 24 hrs and its preparation completed as described in regard to S–I above. In the following, the Sepharose-bound insulin antibodies are referred to as S-AIS.

The binding capacity of the S-AIS was determined by adding increasing amounts of ^{125}I pork insulin, ^{125}I pork proinsulin or ^{125}I ox proinsulin, shaking overnight at 4° C, centrifuging, washing and counting the precipitate. Based on its estimated capacity to bind ^{125}I pork insulin, an S-AIS suspension capable of binding 0.1 U/ml was prepared in NaFAM and stored in 1 ml portions at − 18° C until use.

Removal of Proinsulin + Insulin

Proinsulin + insulin in serum samples or pancreas extracts was bound to S-AIS in the following manner: To 1 ml of serum was added 100 μl of an S-AIS suspension with a binding capacity of 0.01 U/ml. The serum samples were shaken overnight at 4° C, centrifuged, and C-peptide determined in the supernatant.

Immunoassay Procedure

To triplicates 100 μl volumes of standard solutions containing 0.05–1.0 pmole human synthetic C-peptide (a gift from Dr. K. Naithani, Wollforschungsinstitut, Aachen, West Germany) per ml was added 100 μl of dilute antiserum, e.g., M 1017 diluted 1:800. The mixture was incubated at 4° C for 24 hours, 100 μl of ^{125}I-Tyr-C-peptide (0.67 nM) was added, the 4° C-24-h incubation repeated and the antibody-bound C-peptide separated using 1.6 ml 95% ethanol, was described for glucagon [5].

Results

Production of Antibodies against Human C-Peptide and Their Use in Preparing Standard Curves

Nine out of the twelve guinea pigs developed antibodies to C-peptide that could be used for immunoassay in final dilutions of 1:1500 or higher. Fig. 1 shows the standard curves of M 1017 (1:2400) after 6 months, M 1181 (1:3000) after 8 months, M 1183 (1:1800) after 6 months and M 1187 (1:3000) after 6 months of immunization – final dilutions in parentheses. The nonspecific coprecipitation (absence of antibodies) by ethanol was in the range of 1–3% of the ^{125}I-Tyr-C-peptide; a surplus of antibodies bound 72–66% of the tracer. However, with a recently improved quality of the tracer, 80% binding was found [12].

Incubation Time and Temperature

A study of the rate of reaction at 4° C between ^{125}I-Tyr-C-peptide and M 1017 showed that it was necessary to incubate for over 10 hours to reach a plateau, and for practical reasons incubations were run overnight.

When ^{125}I-Tyr-C-peptide was incubated with dilute M 1017 at 4 and 25° C for 24 hrs (10 tubes at each temperature) the mean binding percentages were 28.4 ± 0.38 and 23.3 ± 0.30, respectively (mean of 10 ± 1 SD). Thus the equilibrium constant appears to be

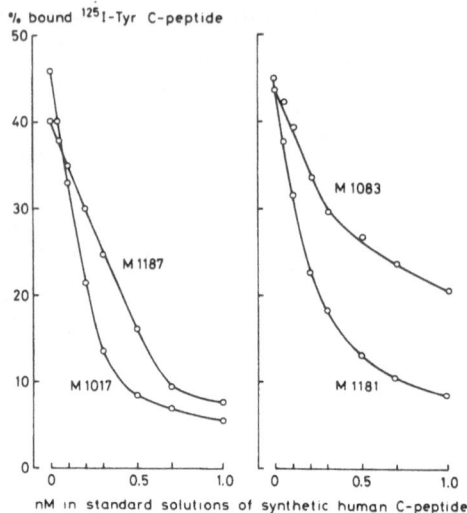

Fig. 1. Standard curves obtained with four different anti-human-b-component guinea pig sera: M 1017 (1:2400), M 1187 (1:3000), M 1083 (1:1800) and M 1181 (1:3000). The final dilution is shown in brackets

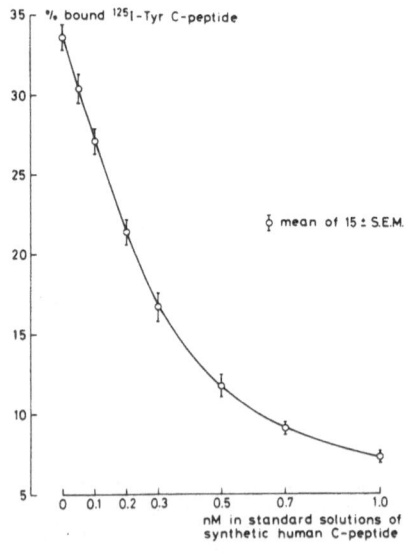

Fig. 2. Reproducibility of the standard curve, using M 1017

higher at 4° C than at 24° C, and all incubations were run at 4° C, ^{125}I-Tyr-C-peptide was added at 4° C, and in the final step the ethanol, too, was added at 4° C.

Reproducibility of the Standard Curve, Detection Limit and Sensitivity

Fig. 2 shows the reproducibility of the standard curve obtained with M 1017 over a period of 7 months, using 3 different batches of ^{125}I-Tyr-C-peptide. Sera from two juvenile diabetics who had been insulin treated for several years, both of whom showed a complete absence of B-cell activity in response to 1 mg of glucagon i.v. [13], were analyzed with 6 of the antisera. The values were between 0 and 0.16 nM. Only one serum, M 1181, gave values not significantly different from zero, while M 1017 gave 0.03–0.07 nM. The detection limit for M 1181 and M 1017 − defined as the smallest quantity of C-peptide that can be distinguished as significantly different from zero point (= 2 SD) − was slightly below 0.03 nM or 0.003 pmole in 100 μl. However, since serum contains substances that do react with some antisera, the true detection limit is higher, except for M 1181. Thus only values higher than 0.07 pmole/ml should be regarded as positive with M 1017. Three standard sera were subjected to repeated assays using both M 1017 and M 1181. The mean values ± 1 SD were: M 1017: (A) 0.58 ± 0.12 (13), (B) 0.91 ± 0.11 (30), (C) 3.53 ± 0.37 (11); and M 1181: (A) 0.45 ± 0.03 (18), (B) 0.72 ± 0.03 (19), (C) 3.27 ± 0.11 (19). The figures in parentheses are the number of assays run with the sera.

In the range of 0–0.3 nM, the decrease in % bound tracer was about 17 (later improved to approximately 30%, see Fig. 1), corresponding to 5.6% per 0.1 nM. As the mean standard deviation of 115 triplicates was 1.87 (relative) per cent, or about 0.6 absolute per cent, the sensitivity defined as 2 standard deviations in the aforementioned range was about 0.0025 pmole in 100 μl for both M 1017 and M 1181.

Preparation of Insulin Antibodies Coupled to Sepharose (S-AIS), Yield and Capacity to Bind Insulin and Proinsulin

The insulins coupled to Sepharose, S-I$_{ox}$ and S-I$_{pork}$, were suspended in phosphate buffer to a calculated concentration of 1 mg bound insulin/ml (based on absorbance). The IRI was estimated to be 4.0–4.9 U/ml for the S-I$_{ox}$, and 3.8–4.1 U/ml for S-I$_{pork}$, meaning that about 20% of the bound insulin exhibited immunoreactivity.

Insulin antibodies were bound to S-I$_{pork}$ at neutral pH and eluted, after 3 washes, with dilute HCl. More than 95% of the antibodies were bound to S–I and the yield of antibodies in the acid eluates was in the range of 25–37%. The coupling procedure was effective, analyses showing that more than 95% of the antibodies had been bound to Sepharose. Table 1 shows the binding capacity of the S-AIS towards ^{125}I-pork insulin and ^{125}I-pork and ox proinsulin (about 750 pmoles/ml). Apparently the S-AIS had less affinity to the proinsulin, since a higher surplus of ^{125}I proinsulin was necessary to occupy all the binding sites as compared to insulin.

Table 1. *Binding capacity of S-AIS toward ^{125}I pork insulin, ^{125}I pork proinsulin and ^{125}I ox proinsulin*

| added ^{125}I insulin | bound ^{125}I insulin per 50 μl S-AIS susp. (1:10) | added ^{125}I pork proinsulin | bound ^{125}I pork proinsulin per 50 μl S-AIS susp. (1:10) | added ^{125}I ox proinsulin | bound ^{125}I ox proinsulin per 50 μl S-AIS suspension (1:10) |
pmoles	pmoles	pmoles	pmoles	pmoles	pmoles
8.4	3.3	4.6	2.7[a]	4.6	2.3[a]
12.6	3.8	6.9	3.1[a]	6.9	2.8[a]
16.8	3.6	9.4	3.6	9.4	3.2[a]
21.0	3.4	11.7	3.5	11.7	3.9
25.2	3.7	14.1	3.9	14.1	3.9
29.4	3.8	16.4	3.8	16.4	3.9
33.8	3.3	18.7	3.9	18.7	4.1
mean ± 1 SD pmole/50 μl (1:10)	3.56 ± 0.22		3.74 ± 0.18		3.95 ± 0.10
pmole/ml	712 (~ 0.1 U/ml)		748		790

[a] not included in the calculation of the mean value

Table 2. *Determination of human C-peptide in a waste product of human insulin production, pmole/ml x dilution. M 1017*

Dilution	Batch 1	Batch 2	Batch 3
1:4000	1520	1680	1480
1:6000	1920	2100	1920
1:8000	2160	2400	2080
1:10000	2300	2400	2000

In the routine separation of insulin + proinsulin from C-peptide, the amount of S-AIS has to be adequate to bind all the insulin sites. 100 µl of an S-AIS suspension with a binding capacity of 0.01 U/ml bound about 95% of up to 10 ng of either ^{125}I ox proinsulin, ^{125}I pork proinsulin or ^{125}I pork insulin. When use was made of solutions of unlabelled human insulin, pork proinsulin and ox proinsulin, similar results were recorded with up to 10 ng/ml, inasmuch as about 95% of the immunoreactivity was removed from 1 ml using 100 µl S-AIS (0.01 U/ml). Binding of human proinsulin to S-AIS was also established employing quantities of 3 and 5 ng/ml (10 ng was not available).

Assay of Pancreatic Extracts Containing C-Peptide

Three batches of mother liquor from the salting-out process (15% NaCl) in the pilot-plant-scale preparation of human insulin were analyzed with a view to comparing the immunoreactivity of the natural C-peptide (or C-peptide-like components) with the synthetic standard, since no purified natural C-peptide was available. From Table 2 it appears that the natural C-peptide consistently showed a dilution pattern slightly different to that of the synthetic standard.

Determination of C-peptide in Serum

Table 3 shows the effect of diluting samples from normal persons with high (> 1 pmole/ml) C-peptide concentrations using M 1017. In most sera a dilution

effect was observed, i.e. higher values were recorded with higher dilutions, as in the case of the pancreatic extracts. With M 1181 the dilution effect was less pronounced.

Table 4 shows the corresponding IRI and C-peptide values measured in 15 normal persons, 8 maturity-onset diabetics and 10 insulin requiring diabetics. None of the diabetics had ever been treated with insulin and therefore had no insulin antibodies. C-peptide was estimated with M 1017 and M 1181.

Discussion

Antibody Production, Incubation Time and Temperature

It was possible to obtain antibodies to human C-peptide in 9 of 12 guinea pigs immunized with human b component. There is no means of comparing this result with other methods used to produce antibodies because there is no indication of the number of positive animals in the relevant publications [2, 8, 14]. However, the technique is still far from satisfactory and other methods, including the coupling of C-peptide, should be investigated when sufficient material becomes available.

The reaction between ^{125}I-Tyr-C-peptide and the antisera had to be carried out at 4° C in order to obtain the maximum of binding and the incubation period had to be extended to over 10 hours.

Preparation of S-I and S-AIS and Their Immunological Activities

The IRI of the insulin covalently bound to Sepharose was about 20%. This probably means that 80% of the insulin is attached to the Sepharose in such a way as to make the immunogenic sites inaccessible to the antibodies. Nevertheless, the S-I is capable of binding

Table 3. *C-peptide determined at different dilution ratios in serum samples from normal persons after oral glucose. Concentration in diluted sample x dilution factor, nM. Antiserum: M 1017*

Dilution	A	B	C	D	E	F	G	H	I	J	K	L	M	N	O	P	Q	R	S	T
1:2	2.0	2.3	1.9	2.0	2.0	2.0														
1:4	2.0	2.2	1.9	2.0	2.0	1.9	2.8	2.2	1.9	1.7										
1:5											1.05	2.05	2.60	2.0	4.2	2.9	3.7	3.4	4.8	
1:6	2.2	2.2	1.9	2.3	2.3	1.9	–													
1:8	2.3	2.0	1.8	2.4	1.3	1.7	2.6	2.4	1.8	1.7										3.9
1:10	2.6	2.3	2.3	2.8	2.3	2.1					1.20	2.30	3.20	2.2	4.2	3.6	3.7	3.8		4.3
1:20													4.0						5.2	4.0

52

Table 4. *IRI and C-peptide in normal subjects, maturity-onset diabetics and insulin-requiring diabetics, fasting and 1 h after 1.75 g oral glucose/kg, in nM (mean ± 1 SD)*. C-peptide determined with M 1017 and M 1181

Group		fasting			1 h after oral glucose		
		IRI	C-peptide		IRI	C-peptide	
			M 1017	M 1181		M 1017	M 1181
15 normal subjects	mean	0.048	0.37	0.35	0.52	2.53	2.24
	SD	0.033	0.071	0.09	0.29	0.70	0.71
8 maturity-onset	mean	0.11	0.86	0.74	0.49	2.49	2.34
diabetics	SD	0.088	0.51	0.51	0.32	0.93	1.18
10 insulin-	mean	0.063	0.37	0.21	0.105	0.49	0.42
dependent diabetics	SD	0.027	0.14	0.14	0.042	0.17	0.24

all the insulin antibodies of guinea pig sera at neutral pH. The elution of antibodies at low pH caused considerable losses — only 25–37% was recovered. It is not known whether the antibodies were destroyed at the low pH or remained bound to the S-I. This is being investigated using ^{125}I-labelled antibodies. Surprisingly, only 50% of the immunoreactivity of the antibodies was lost in the coupling to Sepharose, so that the overall yield of S-AIS was about 15%. The binding capacity of the S-AIS was the same towards proinsulin and IRI. S-AIS with a binding capacity of 1 mU was sufficient to remove 95% of IRI in 1 ml volumes containing up to 10 ng of either insulin or proinsulin (both labelled and unlabelled). Therefore, sera having IRI > 10 ng/ml should either be diluted before adding S-AIS or S-AIS should be added in higher amounts. Serum from insulin-treated patients which contains insulin antibodies should not be treated with S-AIS directly. In those patients who have some residual B-cell activity human proinsulin has been shown to be present, bound as it were to the endogenous insulin antibodies [1, 3], and therefore proinsulin will not be removed from their serum by S-AIS. The serum must be subjected to extraction [4] prior to treatment with S-AIS and determination of C-peptide.

Sources of Error in the Determination of C-Peptide in Serum

Synthetic C-peptide was used as the standard because no natural peptide was available. Pancreatic extracts containing approximately 2000 pmole of C-peptide/ml showed, however, a dilution pattern somewhat different to that of the synthetic standard. The pancreatic extracts probably contain a variety of C-peptide fragments, which may be an explanation of the difference. Another possibility is that the synthetic standard is different from the natural C-peptide. A second batch of synthetic human C-peptide was prepared recently [11]; it deviates from the batch employed in this study and has shown considerable similarity to natural C-peptide. A comparative study of the synthetic connecting peptide and the natural C-peptide [8], using antibodies against the connecting peptide (C-peptide = connecting peptide lacking two pairs of basic amino acids at each terminal), concludes that the two peptides have approximately the same degree of reactivity, through without showing the results. In another paper [17], synthetic C-peptide was reported to have 85–90% of the immunoreactivity of the natural C-peptide with an antiserum against natural C-peptide. Table 3 shows the dilutions of a series of samples with C-peptide levels higher than 1 pmole/ml. Dilution effects similar to the one observed in pancreatic extracts were observed in many samples. Again, the difference in reactivity could be due to the presence of degraded fragments of C-peptide, or it could represent a difference in the characteristics of the natural and the synthetic C-peptide.

Another source of error is the displacement of the tracer, which was first observed with sera from long-term insulin treated juvenile diabetics who had shown no C-peptide or IRI response to i.v. glucagon. The displacement varies from one antiserum to another (0–0.16 nM) and the fact that zero values were consistently found with M 1181 shows that this phenomenon is most likely due to substances other than C-peptide and proinsulin. This is supported by the finding of a weak displacement with pure human IgG (own observation). Moreover, determination of C-peptide in sera devoid of insulin antibodies with M 1017 (up to 0.07 nM in the "zero serum") and M 1181 revealed slightly lower values with M 1181 (Table 4). During suppression of endogenous insulin by fish insulin in normal persons [15] it was observed that C-peptide dropped in some cases to undetectable values after 3–4 hours when measured with M 1181, whereas the values recorded with M 1017 were approximately 0.07 nM in the same samples [7]. All the aforementioned findings lend support to the hypothesis that some antisera to human C-peptide are capable of reacting, to varying degrees, with substances in serum that are different from C-peptide, proinsulin and fragments thereof. It is interesting to note that the SD of repeated determinations of C-peptide in different assays in the same control sera was about 4 times less with M 1181 than with M 1017 (see page 8). Although M 1017 and M 1181 yield practically identical standard curves, M 1181 was chosen for the determination of C-peptide due to its "zero value" and its low assay-to-assay variation.

The mean C-peptide concentrations registered in 15 normal persons using M 1181 — basal and 1 h after oral glucose (1.75 g/kg) — were 0.35 ± 0.1 nM ($= 1.1 \pm 0.3$ ng/ml) and 2.24 ± 0.71 nM ($= 6.7 \pm 2.1$ ng/ml). The basal value was slightly lower than the value of 1.3 ± 0.3 ng/ml reported by [2] and slightly higher than 0.88 ± 0.21 as reported by [8]. The recordings after glucose show greater variations, from 6.7 ± 2.1 ng/ml found here to 4.4 ± 0.8 [2] and 5.1 ± 0.7 ng/ml [8]. Differences in the standards used are a likely explanation, though the fact that serum was added to the standards in [2] and [8] may have reduced the values recorded in these assays. This will show up most clearly in samples with high C-peptide concentrations, which have to be diluted.

Methodology in General

As C-peptide has no biological activity, the radioimmunoassay is the only analytical technique available for its determination. In spite of the interest currently focussed on C-peptide, only three research groups have succeeded in establishing an assay [2, 6, 8] due to the numerous problems it presents, such as: preparation of C-peptide and ^{125}I-Tyr-C-peptide, production of antibodies and separation of C-peptide from proinsulin-containing substances. The first assay to be described [14] used antibodies against natural C-peptide coupled to albumin which had low capacity (final dilution 1:500) and low affinity (detection limit: 0.75 ng total added). In addition to this, separation of C-peptide from PLI by gel filtration of serum extracts probably resulted in heavy losses of C-peptide, for C-peptide and insulin were found to be present in peripheral blood in equimolar amounts.

Later, the gel filtration procedure was left out [2] and total C-peptide immunoreactivity (CPR) was measured in unextracted serum. In this assay, C-peptide was found to be 1.3 ± 0.3 ng/ml in fasting normal persons, i.e., a value much higher than IRI.

In an assay based exclusively on synthetic human connecting peptide [8], use was made of antibodies in a final dilution of 1:17000 and a detection limit of 0.1 ng per tube. No separation was performed, so that the assay was in regard of total CPR.

This paper reports on the production of antibodies using human b-component, which could be used in a final dilution of 1:2100 — 1:3000, with the best serum giving a detection limit of 0.003 pmole in 100 μl. This is a significant feature because many diabetic patients have particularly low C-peptide levels.

The separation technique reported here is very quick and features advantages over the techniques reported earlier [6], where the PLI + IRI was bound to insulin antibodies and removed by ethanol precipitation, necessitating evaporation before analysis. The use of Sepharose-coupled antibodies (S-AIS) for binding of PLI + IRI implies that, following centrifugation, C-peptide can be determined direct in the supernatant. Although total CPR will be very close to C-peptide values in many normal samples, the determination of C-peptide should be preferable to determining total CPR since this assay will be applied primarily to diabetic sera, which — in the case of maturity-onset type diabetics – have been shown to have elevated proinsulin [1]. The same applies to insulinoma patients. Furthermore, as long as we do not know the ratio of proinsulin to C-peptide in, e.g., juvenile diabetics, and being, moreover, cognizant of the fact that proinsulin and C-peptide react differently with the antibodies used so far [2, 6, 8], it is preferable to remove PLI before estimating C-peptide.

In insulin treated patients who have antibodies to insulin it is necessary to include an acid extraction of the serum before adding S-AIS and determining C-peptide.

Finally, it should be stressed that there still are some sources of error in the C-peptide assay that remain to be solved. First of all, there is the problem of identity of synthetic and natural C-peptides and of characterization of the circulating C-peptide, which could be a fragment of the intact molecule, for instance fragment 1–24, which was found in pancreatic extracts [6].

Acknowledgements. The excellent technical assistance of Mrs. Lisbet Pedersen and Miss Marianne Knudsen is gratefully acknowledged.

References

1. Block, M. B., Mako, M., Steiner, D. F., Rubenstein, A. H.: Elevated circulating proinsulin levels in insulin-requiring diabetic patients. J. Lab. clin. Med. **78**, 811–812 (1971)
2. Block, M. B., Mako, M. E., Steiner, D. F., Rubenstein, A. H.: Circulating C-peptide immunoreactivity. Studies in normals and diabetic patients. Diabetes **21**, 1013–1026 (1972)
3. Fink, G., Cresto, J. C., Gutman, R. A., Lavine, R. L., Rubenstein, A. H., Recant, L.: Plasma proinsulin-like material in insulin treated diabetics. Horm. Metab. Res. **6**, 439–443 (1974)
4. Heding, L. G.: Determination of total serum insulin (IRI) in insulin treated diabetic patients. Diabetologia **8**, 260–266 (1972)
5. Heding, L. G.: Radioimmunological determination of pancreatic and gut glucagon in plasma. Diabetologia **7**, 10–19 (1972)
6. Heding, L. G., Larsen, U. D., Markussen, J., Jørgensen, K. H., Hallund, O.: Radioimmunoassays for human, pork and ox C-peptides and related substances. Horm. Metab. Res. Suppl. Vol. **5**, 40–44 (1974)
7. Heding, L. G., Turner, R. C., Harris, E.: C-peptide, proinsulin and insulin responses to fish-insulin induced hypoglycaemia in the diagnosis of insulinomas. Diabetes **24** (Suppl. 2), 412 (1975)
8. Kaneko, T., Oka, H., Munemura, M., Oda, T., Yamashita, K., Suzuki, S., Yanaihara, N., Hashimoto, T., Yanaihara, C.: Radioimmunoassay of human proinsulin C-peptide using synthetic human connecting peptide. Endocr. jap. **21**, 141–145 (1974)
9. Markussen, J., Sundby, F., Smyth, D. G., Ko, A.: Preparation of human C-peptide. Horm. Metab. Res. **3**, 229–232 (1971)
10. Markussen, J., Heding, L. G., Jørgensen, K. H., Sundby, F.: Proinsulin, insulin and C-peptide. Horm. Metab. Res. Suppl. Series No. **3**, 33–35 (1971)
11. Naithani, V. K., Dechesne, M., Markussen, J., Heding, L. G., Larsen, U. D.: Studies on Polypeptides, VI[1], synthesis, circular dichroism and immunological studies of tyrosyl human C-peptide. Hoppe-Seylers Z. physiol. Chem. **356**, 1305–1312 (1975)
12. Naithani, V. K., Dechesne, M., Markussen, J., Heding L. G.: Studies on polypeptides, V[1]. Improved synthesis of human proinsulin C-peptide and its benzyloxy-carbonyl derivative. Circular dichroism and immunological studies of human C-peptide. Hoppe-Seylers Z. physiol. Chem. **356**, 997–1010 (1975)
13. Östman, J., Arner, P., Groth, C.-G., Heding, L., Jorfeldt, L., Lundgren, G.: Pancreastransplantation vid diabetes mellitus: Studier av pankreas endokrina funktion och intermedicinska synspunkter. (paper in preparation)
14. Rubenstein, A. H., Block, M. B., Starr, J., Melani, F., Steiner, D. F.: Proinsulin and C-peptide in blood. Diabetes **21** (Suppl. 2), 661–672 (1972)
15. Turner, R. C., Harris, E.: Diagnosis of insulinomas by suppression tests. Lancet **1974 II**, 188–190
16. Schlichtkrull, J.: Proinsuline et substances apparantées. Path. et Biol. **19**, 885–892 (1971)
17. Yanaihara, N., Hashimoto, T., Yanaihara, C., Sakagami, M., Steiner, D. F., Rubenstein, A. H.: Synthesis of human connecting peptide derivatives and their immunological properties. Biochem. biophys. Res. Commun. **59**, 1124–1130 (1974)

Dr. L. G. Heding
Novo Research Institute
Novo Allé
DK-2880 Bagsvaerd
Denmark

Diabetologia 13, 467–474 (1977)

© by Springer-Verlag 1977

Specific and Direct Radioimmunoassay for Human Proinsulin in Serum

L. G. Heding

Novo Research Institute, Copenhagen, Denmark

Summary. A routine radioimmunoassay for human proinsulin in serum has been developed. The reagents used were: antibodies against the C-peptide part of the proinsulin molecule, human proinsulin as the standard and ^{125}I-labelled synthetic human Tyr-C-peptide as the tracer. The first step in this assay comprises the binding of proinsulin to insulin antibodies covalently coupled to Sepharose (S-AIS). Although bound to the solid-phase S-AIS, the proinsulin retains its second immunogenic site, viz., the C-peptide part of the molecule, accessible. Hence a surplus of C-peptide antibodies is added to the S-AIS-bound proinsulin, and the residual amount of C-peptide antibody is determined by addition of ^{125}I-Tyr-C-peptide. The detection limit is approximately 0.01 pmol/ml. The advantages of this method are: (1) its high specificity (proinsulin is determined as a molecule having both an insulin and a C-peptide moiety), (2) its simplicity and rapid performance, and (3) the low detection limit of the assay. Fasting sera from 24 nondiabetics, 9 maturity-onset diabetics and 10 newly diagnosed insulin requiring diabetics showed the following concentrations of proinsulin: 0.009 ± 0.005, 0.022 ± 0.23 and 0.010 ± 0.009 pmol/ml (mean ± SD). One hour after 1.75 g/kg oral glucose, the values increased to 0.052 ± 0.023, 0.046 ± 0.022 and 0.032 ± 0.022 pmol/ml. The fasting proinsulin constituted 19, 23 and 17% of the IRI, respectively, whereas 1 h post glucose these values changed to 8, 9 and 31% of IRI. Serum from 10 insulin-treated diabetics containing insulin antibodies contained from 0–1.80 pmol/ml, whereas the C-peptide levels in the same patients were 0–0.35 pmol/ml. It is suggested that insulin requiring diabetics hypersecrete proinsulin due to the inability of their B-cell to arrange proinsulin in secretory granules for adequate proinsulin/insulin conversion.

Key words: Radioimmunoassay, human proinsulin, insulin, C-peptide, oral glucose, diabetics.

Proinsulin, the biosynthetic precursor of insulin, was discovered by Steiner et al. [1]. The initial methods of quantitative determination of proinsulin in serum, which also contains insulin and C-peptide, have been based on (i) proinsulin having a higher molecular weight (9000) than insulin (6000), and (ii) its reaction with insulin antibodies. Thus, when serum or an extract of serum was subjected to gel filtration on Sephadex G 50 [2] or Biogel P 30 [3] two peaks of insulin immunoreactivity could be distinguished, one corresponding to insulin and the other one having a higher molecular weight. The latter was referred to either as "big" insulin [2] or the proinsulin-like component [4]. In order to characterize further the high-molecular-weight immunoreactive insulin (IRI) in the fractionated serum, each fraction was analyzed using the human C-peptide assay [3] and C-peptide immunoreactivity was detected in both IRI containing peaks. This strongly indicated that the first peak contained proinsulin, which exerts both insulin and C-peptide immunoreactivity. Furthermore, it was shown that C-peptide eluted in the second peak together with insulin in spite of its low molecular weight (approx. 3000).

The gel filtration procedure is, however, time consuming and has been reported to require at least 60–100 µU [5] or 4–5 ml of serum when the insulin concentration was less than 20 µU/ml [4]. A new principle of separation was introduced in degrading insulin in serum with an insulin specific protease whilst leaving proinsulin unaltered and detectable with the insulin radioimmunoassay [6].

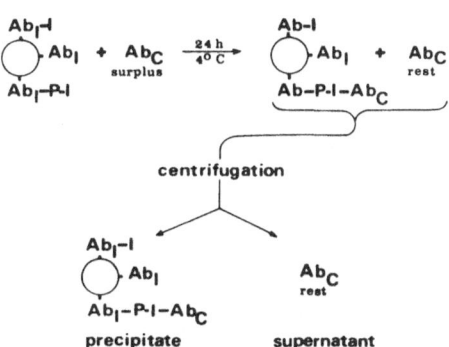

(1)

Ab$_I$
◯–Ab$_I$ + I + P–I + C $\xrightarrow[4°C]{24 h}$ ◯–Ab$_I$–I + C
Ab$_I$ Ab$_I$–P–I

centrifugation

Ab$_I$–I
◯–Ab$_I$ C
Ab$_I$–P–I
precipitate   supernatant

(2)

Ab$_I$–I
◯–Ab$_I$ + Ab$_C$ $\xrightarrow[4°C]{24 h}$ ◯–Ab$_I$ + Ab$_C$
Ab$_I$–P–I   surplus   Ab–P–I–Ab$_C$ rest

centrifugation

Ab$_I$–I
◯–Ab$_I$ Ab$_C$
Ab$_I$–P–I–Ab$_C$ rest
precipitate   supernatant

Fig. 1. Principle of the direct proinsulin radioimmunoassay. Diagram 1 shows the separation procedure using S-AIS, diagram 2 shows the principle of the assay where the S-AIS bound proinsulin is capable of binding antibodies to C-peptide. Ab, insulin antibodies; I, insulin; P–I, proinsulin; C, C-peptide; Ab$_C$, C-peptide antibodies.

The purpose of this paper is to present a new specific and direct radioimmunoassay for proinsulin in serum which allows the assay of a large number of samples simultaneously [8].

Materials and Methods

The principle of the assay is shown in Figure 1. Insulin antibodies covalently coupled to Sepharose are added to serum containing insulin, proinsulin and C-peptide. The solid-phase insulin antibodies bind any molecule containing an insulin moiety, in this case

insulin and proinsulin. The C-peptide remains unbound in the supernatant and is determined by radioimmunoassay without interference from proinsulin [9]. The amount of bound proinsulin is then measured directly by the C-peptide assay since it has retained its ability to bind antibodies to proinsulin.

Human C-peptide in serum was determined according to [9], using anti-proinsulin serum from guinea pig M 1181, ^{125}I-labelled synthetic Tyr-C-peptide [10] as the tracer, and synthetic C-peptide [11] as the standard.

The IRI of serum not containing insulin antibodies was determined by direct radioimmunoassay [12]. Serum containing antibodies against insulin was extracted at low pH prior to the determination of IRI [12]. Insulin antibodies raised in guinea pigs and purified by reaction with monocomponent insulin coupled to Sepharose, were coupled to Sepharose 4B (referred to as S-AIS) as described in [9]. The binding capacity of the S-AIS was determined by adding increasing amounts of ^{125}I-labelled tracers of porcine insulin, porcine proinsulin and bovine proinsulin. A stock suspension of S-AIS with an estimated binding capacity of 0.1 U/ml was stored in 1 ml portions at −18° C until use [9].

The proinsulin assay procedure was as follows: To 1 ml of serum or standard solution of human proinsulin (0.02–0.2 pmol/ml ∼ 0.18–1.8 ng/ml), kindly donated by Professor Arthur Rubenstein of Chicago, was added 100 µl of S-AIS suspension with a binding capacity of 0.01 U/ml. The samples were shaken at 4° C overnight, centrifuged, and 900 µl of the supernatant removed using an Oxford Sampler Model Q. The S-AIS containing bound insulin and proinsulin was then washed twice with 2 ml of phosphate buffer (0.04 mol/l, pH 7.4) containing human albumin (Behringwerke, electrophoretic purity 100%) (60 g/l) and thiomersal (0.2 g/l), in the following referred to as NaFAM, in order to remove the residual C-peptide (top diagram in Figure 1). After each wash, 2.0 ml was pipetted off and discarded, so that 200 µl remained in each tube. After the second wash, 300 µl of NaFAM and 200 µl of anti-proinsulin serum M 1181 (1:700) [9] were added and the tubes shaken at 4° C overnight (bottom diagram in Figure 1). After centrifugation, 2 × 200 µl were pipetted from each supernatant into two glass tubes (100 × 10 mm) and 100 µl of ^{125}I-Tyr-C-peptide (0.33 pmol/ml = 1 ng/ml) was added. The mixture was incubated at 4° C for a further 24 h and the antibody-bound ^{125}I-Tyr-C-peptide was separated by addition of 1.6 ml 95% (v/v) ethanol, as described for glucagon [13].

The concentrations of insulin, C-peptide and proinsulin are given in pmol/ml. 1 µU of insulin

= 0.0062 pmol/ml (27 U/mg of MC human insulin), 1 pmol/ml of insulin = 6 ng/ml, 1 pmol/ml of C-peptide = 3 ng/ml, and 1 pmol/ml of proinsulin = 9 ng/ml. Due to the shortage of human proinsulin the number of standards as well as the number of recovery experiments were limited to the least necessary.

To extracts of serum from insulin treated diabetics whose IRI ranged 200–2000 μU/ml, was added 100 μl of S-AIS with a binding capacity of 0.1 U/ml. Whenever the IRI exceeded 2000 μU/ml, the extracts were diluted prior to the addition of S-AIS.

Insulin binding to IgG was determined by the immunoelectrophoretic method of Christiansen [39].

Porcine and bovine proinsulin were determined as described in [14]. Recovery of proinsulin after the acid extraction was checked using circa 97% pure porcine and bovine proinsulin [14].

Serum samples from 24 non-diabetics (aged 15–65 years, mean: 37 years, % ideal body weight: 87–134, mean: 108) and 21 newly diagnosed diabetics, fasting and 1 h post 1.75 g/kg oral glucose were obtained from Dr. S. Munkgaard Rasmussen of Hvidøre Hospital (presently of Bispebjerg Hospital, Copenhagen). These diabetics had been divided into two groups: 11 maturity-onset diabetics (aged 24–72 years, mean age: 46 years, % ideal body weight: 85–179, mean: 118) and 10 insulin requiring diabetics (aged 17–70, mean age: 38 years, % ideal body weight: 81–142, mean: 100). None of these patients had ever received insulin and they had no insulin antibodies. Serum samples from 10 juvenile-type diabetics (aged 12–22, mean age: 18 years) treated with conventional insulin for more than one year, all having insulin antibodies, were obtained from Dr. J. Ludvigsson, Department of Paediatrics, University Hospital, Linköping, Sweden, and Dr. Beck Nielsen, Aarhus Amtshospital, Aarhus.

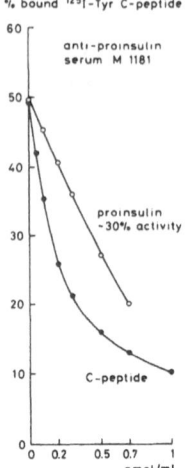

Fig. 2. Standard curves of proinsulin and C-peptide using antiserum M 1181

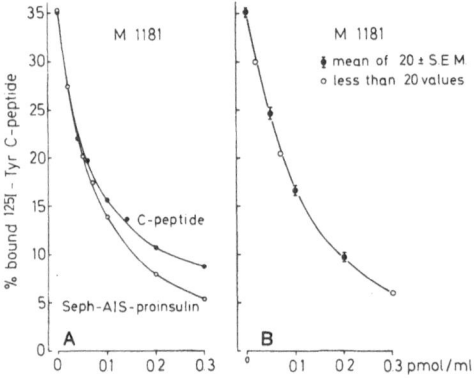

Fig. 3. A Standard curves of S-AIS bound proinsulin (Seph-AIS-proinsulin) and C-peptide. **B** Reproducibility of the 20 S-AIS proinsulin standard curves set up over a period of 12 months

Results

Reactivity of Free and S-AIS-Bound Proinsulin

Figure 2 shows the reaction to antiserum M 1181 of human proinsulin in solution in the C-peptide assay as compared to C-peptide. Three further antisera were tested (M 1017, M 1180 and M 1187), and proinsulin exhibited in all cases much less (30–50%) reactivity than C-peptide. In contrast, S-AIS-bound proinsulin (Fig. 3A) showed approximately the same reactivity as C-peptide with the same antiserum (M 1181).

Reproducibility of the Standard Curve, Detection Limit and Sensitivity

Figure 3B shows the reproducibility of the standard curve obtained with M 1181 over a period of 12 months using 6 different batches of ^{125}I-Tyr-C-peptide. Only those solutions containing 0.05–0.1 and 0.2 pmol/ml were included in the assays, but lately a solution containing 0.02 pmol/ml has also been in use with a view to improving the accuracy in the low range. The detection limit, defined as the smallest quantity of proinsulin that can be distinguished as significantly different from zero (= 2 SD), was

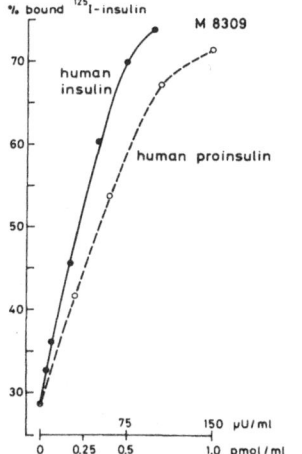

Fig. 4. Standard curves of human insulin and proinsulin in the IRI assay. Antiserum M 8309 was used

slightly below 0.01 pmol/ml; therefore all readings $\geqq$ 0.01 pmol/ml were regarded as positive. In the range of 0–0.1 pmol, the % bound tracer decreased by about 19, corresponding to 1.9% per 0.01 pmol/ml. As two SD in the above-mentioned range were 1% or less, the sensitivity defined as two SD was less than 0.01. Removal of C-peptide was checked in each assay by including a sample containing 2 pmol/ml of C-peptide. This sample could not be distinguished from zero.

Recovery and Dilution of Samples

Addition of 0.02, 0.05 and 0.1 pmol/ml to normal sera gave a virtually 100% recovery. Dilution of serum containing 0.03–20 pmol/ml (normal sera 1 h post OGTT and sera from insulinoma patients) has given the expected values within the limits of the assay (e.g., 1:50 and 1:100 − 7.1 and 7.3 pmol/ml; 1:10 and 1:20 − 0.76 and 0.69; undiluted A: 0.09; 1:2 − 0.12; 1:4 − 0.10; 1:8 − 0.10, and undiluted B: 0.18; 1:2 − 0.20; 1:4 − 0.24, 1:8 − 0.18).

The recovery of proinsulin added to serum after acid extraction [12] was determined using porcine and bovine proinsulin. The recovery of both species varied from 80–88% (mean 85%), as determined by the porcine and bovine proinsulin radioimmunoassays, respectively, and it was assumed that the recovery of human proinsulin after extraction would be the same and the results were corrected accordingly.

Reactivity of Proinsulin in the Insulin Radioimmunoassay

In order to calculate the percentage of the IRI due to proinsulin, the reactivity of proinsulin was determined in the insulin radioimmunoassay. The reactivity of bovine proinsulin with anti-insulin serum M 8309 in the 0–0.25 pmol/ml range (0 − approx. 40 µU/ml) has been shown to be similar (on a molar basis) to that of bovine insulin [16]. At higher concentrations, bovine proinsulin has increasingly lower reactivity. In the same assay, human proinsulin shows approximately 66% reactivity in the range 0–0.50 pmol/ml, that is to say, 0.50 pmol/ml of proinsulin reacts like 50 µU/ml of insulin (Fig. 4). The percentage was then calculated as follows: the proinsulin concentration in pmol/ml was converted into µU/ml by multiplying by 100 and this value was then expressed in per cent of the total IRI.

Degradation of Proinsulin in Serum

^{125}I-labelled bovine proinsulin was added to sera (two normal and two diabetic serum samples) and incubated at 37° C for 24 h. Addition of a specific, non-insulin-binding anti-proinsulin serum [17] revealed that ^{125}I-proinsulin was not degraded to any detectable degree. Serum samples stored in deep-freezers and rethawed several times have been found unchanged in analyses performed over several months.

Proinsulin, C-Peptide and IRI in Non-Diabetics and in Newly Diagnosed Diabetics

Table 1 shows the glucose, proinsulin, C-peptide and IRI levels in fasting sera and in sera drawn 1 h after 1.75 g of oral glucose per kg of ideal bodyweight.

In the non-diabetic group, the IRI increased some 10-fold after glucose (from 0.047 to 0.453 pmol/ml, or from 7.7 to 67.7 µU/ml) whereas the increase in the C-peptide and proinsulin was only about 5-fold.

In the group of maturity-onset diabetics, the increase in IRI was approx. 7-fold while C-peptide showed only a 3-fold increase and proinsulin doubled. There was a distinct deviation from the non-diabetics in the group of insulin requiring diabetics, whose IRI and C-peptide levels hardly doubled while the mean proinsulin value increased 3-fold.

Table 2 shows the proinsulin calculated as per cent of the IRI. In several of the samples from fasting subjects, the IRI and proinsulin were zero (hence the reduced number of subjects as compared to Table 1). It appears that proinsulin consituted on

Table 1. Blood glucose, serum IRI, C-peptide and proinsulin before and 1 h after oral glucose (1.75 g/kg) in non-diabetics and newly diagnosed diabetics

Group (No. of patients)		glucose mmol/l basal	1 h	IRI pmol/ml basal	1 h	C-peptide pmol/ml basal	1 h	proinsulin pmol/ml basal	1 h
Non-diabetics (24)	mean	4.64	7.84	0.048	0.453	0.38	1.67	0.009	0.053
	range	4.07–5.83	4.40–12.21	0–0.111	0.105–0.930	0.21–0.57	0.70–3.00	0–0.024	0.010–0.110
	SD	0.40	1.92	0.030	0.294	0.11	0.55	0.005	0.023
Maturity-onset-type diabetics (11)	mean	7.46	14.77	0.068	0.476	0.54	1.54	0.022	0.046
	range	5.28–12.65	12.15–21.61	0–0.173	0.111–1.972	0.18–1.00	0.90–3.90	0–0.103	0.028–0.099
	SD	2.84	3.43	0.059	0.575	0.31	0.94	0.0342	0.022
Insulin requiring diabetics (10)	mean	11.12	19.02	0.048	0.087	0.24	0.44	0.010	0.032
	range	7.64–17.16	13.20–26.84	0.012–0.087	0.019–0.099	0.07–0.41	0.08–0.90	0–0.028	0–0.083
	SD	3.52	4.63	0.023	0.029	0.12	0.30	0.009	0.007

average 20%, though with a wide range of variation in the fasting subjects in all three groups. This decreased to about 10% in the non-diabetics and maturity-onset diabetics, all of whom showed a considerable increase in IRI after glucose. In contrast, the insulin requiring diabetics, whose mean IRI increased by 6 µU/ml, secreted a higher percentage of proinsulin. Only in one patient did the percent proinsulin remain the same after the glucose load.

Proinsulin in Insulin-Treated Juvenile Diabetics

Table 3 shows the IgG binding of insulin, total extractable IRI, C-peptide and proinsulin levels in sera of 10 fasting juvenile diabetics treated with conventional insulin for more than 1 year. All 10 patients had insulin antibodies and, consequently, fasting total IRI levels that were elevated as compared to those in normal subjects (52 normals had a mean fasting IRI level of 7.5 µU/ml = 0.049 pmol/ml, ranging from 0 to 16, 1 SD = 4.4 µU/ml). The C-peptide in three patients from this small group was within the normal range (52 normal subjects had a mean fasting C-peptide level of 0.35 pmol/ml, ranging from 0.18 to 0.57, SD = 0.10 pmol/ml). However, their human proinsulin levels were much higher than in normal persons (see Table 1).

Discussion

Methodology, Sources of Error

Since the proinsulin molecule contains both an insulin and a C-peptide moiety, proinsulin reacts in the radioimmunoassays both for insulin and for C-peptide. As serum contains insulin, C-peptide and proinsulin a specific assay for proinsulin includes a thorough separation procedure. So far, the assays were based on two completely different principles: (i) Separation according to molecular size by gel filtration followed by IRI determinations on the indi-

Table 2. Proinsulin, as per cent IRI in sera of non-diabetics and diabetics, fasting and 1 h after oral glucose (1.75 g/kg)

Group	Proinsulin as % IRI: mean (n) and range			
	basal		1 h	
Non-diabetics	19 (16)	6–34	8 (24)	3–20
Maturity-onset-type diabetics	23 (5)	11–43	11 (9)	5–24
Insulin requiring diabetics	16 (8)	7–26	31 (9)	10–100

Table 3. Insulin antibodies, total IRI, C-peptide and proinsulin in insulin-treated juvenile diabetics

Case (initials)	Age (years)	IgG binding mU/ml	Total IRI µU/ml	pmol/ml	C-peptide pmol/ml	Proinsulin pmol/ml
LP	20	1.044	422	2.78	0.08	0.150
VKA	20	5.655	1168	7.78	0.35	1.80
JLN	19	2.417	512	3.39	0.13	0.325
LJ	19	0.030	48	0.31	0.20	0.105
AUH	19	0.575	304	2.00	0.08	0.163
UH	22	0.175	64	0.42	0.04	0.355
PA	17	1.188	192	1.27	0.11	1.250
SL	12	3.251	640	4.24	0.05	0.072
CL	17	0.097	80	0.53	0.20	0.375
BH	15	0.508	99	0.65	0.12	0.193

vidual fractions [2, 3, 18]. (ii) Enzymatic degradation of insulin followed by determination of the non-degraded proinsulin by insulin radioimmunoassay [6, 19, 20]. The first method is unsuitable for routine analyses of a large number of samples. Furthermore it should be stressed that employing the insulin radioimmunoassay for measurement of proinsulin is only valid when its reactivity has been determined in that particular radioimmunoassay, as pointed out by [21], who found that proinsulin showed less reactivity in their radioimmunoassay than insulin (about 30%). In the insulin assay used in this study, the reactivity of proinsulin was approx. 66% that of insulin and it gave linear dilutions in the range of 0–0.4 pmol/ml, in contrast to bovine proinsulin, whose reaction in the 0–0.75 pmol/ml range was virtually identical with insulin [16].

The enzymatic degradation method, which is quick and requires only small volumes of serum, seemed an attractive alternative to gel filtration. However, it recorded consistently higher fasting values [5, 7] (e. g., 70% proinsulin in normal women [22]) than those recorded by gel filtration. Lately, it has been demonstrated that the degradation of insulin is only partial and dependent on its concentration, and that proinsulin, too, is degraded to some degree [7]. Furthermore, the activity of the enzyme varied in different plasma samples [5]. In a recent paper published by the group who developed the method, the authors mention that PLM (proinsulin-like material) was determined only in plasma containing more than 30 µU/ml of TIR (total immunoreactive insulin) due to the incomplete degradation of insulin in fasting plasma [37]. Hence all the earlier values of % proinsulin in fasting plasma obtained with this method may be regarded as erroneous. However, after OGTT, PLM was found to constitute 13% in normals and 16% in maturity-onset diabetics [38, 20], which is in fair agreement with the present results, although it should be mentioned that the standard used for PLM quantitation was porcine insulin, implying underestimation of proinsulin. The same applies to some of the results obtained with gel filtration [4].

The method presented here has several advantages over the other two:

a) it is quick and simple, its sensitivity is about 0.01 pmol/ml, corresponding to 1 µU of IRI/ml, the total sample requirement is no more than 1 ml of serum or plasma, and it is suitable for routine processing of a large number of samples;

b) its specificity is ensured by the fact that proinsulin can bind to S-AIS whilst still binding antibodies to C-peptide;

c) the separation step is a preliminary to the assays both for C-peptide and for proinsulin. One ml of serum plus S-AIS provides a supernatant in which C-peptide can be determined without interference from proinsulin, and a precipitate free of C-peptide for the determination of proinsulin;

d) identical treatment of standard solutions of proinsulin and samples.

In agreement with the findings of [21], it was found that unbound proinsulin was less reactive with the antibody M 1181 than C-peptide. But, unexpectedly, proinsulin bound to S-AIS showed practically the same reactivity as C-peptide (Figs. 2 and 3), which rendered the assay more sensitive. No intermediate forms of human proinsulin were available, but it is fair to assume that they would behave very much like proinsulin, as was the case with bovine and porcine intermediates [14].

No degradation of ^{125}I bovine proinsulin was observed in serum allowed to stand at 37° C for 24 h, nor did repeated freezing and thawing affect the values. Hence, proinsulin appears to behave as a stable molecule, like insulin, and its resistance to hepatic degradation seems to be better than that of insulin [23]. Long-term experiments have been initiated to establish the storage properties of proinsulin in serum at −18° C, +4° and +25° C.

Since the content of proinsulin in normal sera is 1–5% of C-peptide on a molar basis (see Table 1) it is of utmost importance that all C-peptide is removed from the S-AIS-bound proinsulin. With this in view, the precipitate is washed twice with 2 ml of buffer, and a control tube containing 2 pmol/ml of C-peptide is included in all assays, and it will not be different from the zero standard. The fact that C-peptide does not undergo any non-specific binding to glass or protein makes it no doubt easier to remove all of C-peptide.

Proinsulin Levels

Proinsulin constitutes about 3% of the IRI in the B-cell [14] and a similar ratio was found in portal venous blood during acute stimulation [18]. However, the mean half-disappearance time for proinsulin in 3 insulinoma patients was about four times longer than for insulin [23], due in part to the lower hepatic extraction of proinsulin [24]. This explains the much higher percentage of proinsulin found in peripheral blood. The values reported here: 0–0.024 pmol/ml or 6–34% of the IRI in fasting normal persons, are in good agreement with the findings of [4], [18] and [25]. After stimulation with oral glucose the concentration of proinsulin increased five-fold but, as IRI increased ten-fold, the mean percentage decreased to about 10 (range: 3–20%), as found by [38, 20] using the enzymatic method and by [4] who separated by gel filtration. [21] reported likewise a drop, though from 40 to 20%, in some four persons. The higher figure is partly due to the fact that the calculations were in regard to per cent weight as against per cent molar weight in this report.

Results published about proinsulin in diabetic patients are but few and these publications are especially sparse in regard to the insulin-requiring diabetic. Tables 1 and 2 show that the 9 maturity-onset diabetics (four of whom had ideal body weight

< 110%) had similar values and percentages of proinsulin as the normals. The four obese subjects had no higher mean proinsulin than the rest of the group. These results closely resemble the findings in a group of 7 patients with chemical diabetes reported by [26], as well as the results obtained 1 h after glucose [38]. These groups of diabetics had also higher than normal IRI both in the fasting state and 1 h after oral glucose. Thus, although the secretion of IRI was insufficient to maintain normoglycaemia and glucose tolerance, no striking deviations from the normal peripheral proinsulin levels were found in this type of diabetic. In contrast to the findings reported here and in [26], another study reports no increase in IRP (immunoreactive proinsulin) during OGTT either in 10 normal children or in 11 patients with chemical diabetes, despite massive increments in IRI [27].

There are few reports on serum proinsulin in newly diagnosed insulin-requiring diabetics, who show extremely little, if any, increase in IRI after OGTT at the time of diagnosis. In a group of nine pregnant severe diabetics the percentage of proinsulin determined after gel filtration was higher (19%) than in pregnant mild diabetics (7.6%) or normals (7.4%); the mean IRI increment in the severe diabetics was 27 µU/ml 1 h after oral glucose in contrast to 103 and 160 µU/ml in the two other groups [35]. In three diabetics having less than 25 µU/ml 1 h after oral glucose, the proinsulin component constituted about 37% in the fasting state and was practically unchanged at 2 h, all values being clearly higher than in normal subjects [36]. In the insulin requiring newly diagnosed diabetics investigated in this paper the increase in IRI after oral glucose was extremely small (from a mean value of 0.048 to 0.087 pmol/ml). In the fasting state, the proinsulin levels (ranging 7–26% of the IRI) were similar to those in normals. After oral glucose, the mean percentage of proinsulin increased to 31 (ranging 10–100%) of the IRI. Hence, it appears that the already stimulated B-cell of the fasting insulin requiring diabetic (mean fasting glucose, 11.1 mmol/l) responds to further stimulation with oral glucose with a far higher proportion of the IRI as proinsulin.

Track et al. [28] have shown that degranulated rat islets secrete higher percentages of proinsulin than granulated islets, and state that if proinsulin is not packaged in secretory granules then only a small portion of the proinsulin will be converted to insulin. The findings reported here in regard to juvenile diabetics may thus be a consequence of the complete absence of granules [29] observed with the use of histological techniques.

In a group of 69 insulin treated juvenile diabetics, detectable C-peptide levels were found in 23% [30], the values varying from 0.04 to 0.60 pmol/ml and clearly indicating residual B-cell activity. The concentration of human proinsulin was very high in a small group of such juvenile insulin-treated diabetics, the values exceeding in most cases those of C-peptide (Table 3). Human proinsulin was also estimated by [31] using the C-peptide assay on gel filtered extracts of serum of 6 insulin treated patients and it was found to vary from 4 to 16 ng/ml, a range similar to the one reported here. Due to the presence of insulin antibodies in the sera of all patients except LJ (Table 3), the levels of proinsulin in these patients were most likely higher than in those whose sera contained no antibodies, because of the prolonged half-life of proinsulin-antibody complexes. Assuming that the affinity of human proinsulin to the antibodies equals that of human insulin, the half-life of human insulin will also be prolonged due to the antibodies. If the ratio of secretion of insulin to proinsulin was normal (100:approx. 3 [18]) then the total IRI would have amounted to 20 times the concentration of proinsulin, contrary to what has been established, leaving out of consideration the injected insulin which contributes to increased total IRI. However, when two diabetics with antibodies were injected with ^{125}I-labelled human proinsulin, the half-life of the tracer was found to be 31 h, in contrast to a half-life of 1.9 h of ^{125}I-insulin [31]. The latter figure was obtained from one patient only and, considering the great individual variation in antibody characteristics [34], one has to conclude that clarification of this problem has to await the more thorough studies of the kind introduced by [31].

Thus it seems likely that the B-cell continues to hypersecrete proinsulin in insulin requiring diabetics, possibly due to its inability to arrange the proinsulin in secretory granules for adequate proinsulin/insulin conversion [28], as suggested by [36]. Insulin antibodies promote degranulation [32, 33], thereby counteracting the granulation and conversion mechanisms.

Acknowledgements. I wish to thank Dr. J. Schlichtkrull for his constructive criticism and invaluable advice, and Mrs. Marianne Heyden, Mrs. Lisbet Pedersen and Mrs. Bente Hansen for their skilled technical assistance.

References

1. Steiner, D.F., Cunningham, D., Spigelman, L., Aten, B.: Insulin biosynthesis: Evidence for a precursor. Science **157**, 697–700 (1967)
2. Gorden, P., Roth, J.: Plasma insulin: Fluctuations in the "big" insulin component in man after glucose and other stimuli. J. Clin. Invest. **48**, 2225–2234 (1969)

3. Melani, F., Rubenstein, A. H., Oyer, P. E., Steiner, D. F.: Identification of proinsulin and C-peptide in human serum by a specific immunoassay. Proc. Natl. Acad. Sci. USA 67, 148–155 (1970)
4. Gorden, P., Sherman, B., Roth, J.: Proinsulin-like component of circulating insulin in the basal state and in patients and hamsters with islet cell tumors. J. Clin. Invest. 50, 2113–2122 (1971)
5. Cresto, J. C., Lavine, R. L., Fink, G., Recant, L.: Plasma proinsulin. Comparison of insulin specific protease and gel filtration assays. Diabetes 23, 505–511 (1974)
6. Kitabchi, A. E., Duckworth, W. C., Brush, J. S., Heinemann, M.: Direct measurement of proinsulin in human plasma by the use of an insulin-degrading enzyme. J. Clin. Invest. 50, 1792–1799 (1971)
7. Starr, J. I., Juhn, D. D., Rubenstein, A. H., Kitabchi, A. E.: Degradation of insulin in serum by insulin-specific protease. J. Lab. Clin. Med. 86, 631–637 (1975)
8. Heding, L. G., Rasmussen, S. M.: Direct and specific determination of human proinsulin in normal and diabetic serum. Diabetologia 12, 396 (1976)
9. Heding, L. G.: Radioimmunological determination of human C-peptide in serum. Diabetologia 11, 541–548 (1975)
10. Naithani, V. K., Dechesne, M., Markussen, J., Heding, L. G., Larsen, U. D.: Synthesis, circular dichroism and immunological studies of Tyrosyl C-peptide of human proinsulin. Hoppe Seylers Z. Physiol. Chem. 356, 1305–1312 (1975)
11. Naithani, V. K., Dechesne, M., Markussen, J., Heding, L. G.: Improved synthesis of human proinsulin C-peptide and its benzyloxycarbonyl derivative. Circular dichroism and immunological studies of human C-peptide. Hoppe Seylers Z. Physiol. Chem. 356, 997–1010 (1975)
12. Heding, L. G.: Determination of total serum insulin (IRI) in insulin-treated diabetic patients. Diabetologia 8, 260–266 (1972)
13. Heding, L. G.: Radioimmunological determination of pancreatic and gut glucagon in plasma. Diabetologia 7, 10–19 (1971)
14. Heding, L. G., Larsen, U. D., Markussen, J., Jørgensen, K. H., Hallund, O.: Radioimmunoassays for human, pork and ox C-peptides and related substances. Horm. Metab. Res. (Suppl.) 5, 40–44 (1974)
15. Heding, L. G., Rasmussen, S. M.: Human C-peptide in normal and diabetic subjects. Diabetologia 11, 201–206 (1975)
16. Markussen, J., Heding, L. G.: Separation of the two double-chain bovine intermediates of the proinsulin-insulin conversion. Chemical, immunochemical, circular dichroism and biological characterization. Int. J. Ppt. Protein Res. 8, 597–607 (1976)
17. Schlichtkrull, J.: Proinsuline et substances apparentées. Pathol. Biol. (Paris) 19, 885–892 (1971)
18. Horwitz, D. L., Starr, J. I., Mako, M. E., Blackard, W. G., Rubenstein, A. H.: Proinsulin, insulin and C-peptide concentrations in human portal and peripheral blood. J. Clin. Invest. 55, 1278–1283 (1975)
19. Michaelis, D., Neumann, I., Schulz, B., Nowak, W., Köcher, W., Michael, R., Heinke, P., Reding, R.: Relationen zwischen poratal- und peripher-venösen Insulin- und Proinsulinkonzentrationen unter Glukoseinfusion beim Menschen. Endokrinologie 64, 257–268 (1975)
20. Hausmann, L., Klimek, J., Kamps, K.: Radioimmunologische Proinsulin-Bestimmung im Serum mit einer "insulinabbauenden Protease" aus Rattenleber. Endokrinologie 66, 56–66 (1975)
21. Rubenstein, A. H., Block, M. B., Starr, J., Melani, F., Steiner, D. F.: Proinsulin and C-peptide in blood. Diabetes 21, (Suppl. 2) 661–672 (1972)
22. Hausmann, L., Klähn, D., Klimek, J., Kaffarnik, H.: Basale und reaktive Proinsulin- und Insulinsekretion bei Frauen mit Übergewicht. Dtsch. med. Wochenschr. 100, 284–292 (1975)
23. Starr, J. I., Rubenstein, A. H.: Metabolism of endogenous proinsulin and insulin in man. J. Clin. Endocrinol. Metab. 38, 305–308 (1974)
24. Rubenstein, A. H., Pottenger, L. A., Mako, M., Getz, G. S., Steiner, D. F.: The metabolism of proinsulin and insulin by the liver. J. Clin. Invest. 51, 912–921 (1972)
25. Gorden, P., Roth, J., Hendricks, C. M., Kahn, C. R.: Plasma proinsulin-like components. Isr. J. Med. Sci. 10, 1212–1221 (1974)
26. Rosenbloom, A. L., Starr, J. I., Juhn, D., Rubenstein, A. H.: Serum proinsulin in children and adolescents with chemical diabetes. Diabetes 24, 753–757 (1975)
27. Burghen, G. A., Etteldorf, J. N., Trouy, R. L., Kitabchi, A. E., Keslensky, S.: Insulin and proinsulin in normal and chemical diabetic children. J. Pediatr. 89, 48–53 (1976)
28. Track, N. S., Frerichs, H., Creutzfeldt, W.: Release of newly synthesized proinsulin and insulin from granulated and degranulated isolated rat pancreatic islets. The effect of high glucose concentration. Horm. Metab. Res. (Suppl.) 5, 97–103 (1974)
29. Gepts, W.: Pathologic anatomy of the pancreas in juvenile diabetes mellitus. Diabetes 14, 619–633 (1965)
30. Ludvigsson, J., Heding, L. G., Leander, E.: Fasting C-peptide in juvenile diabetics beyond the remission period in relation to clinical manifestations and treatment at onset. Diabetologia 12, 407 (1976)
31. Fink, G., Cresto, J. C., Gutman, R. A., Lavine, R. L., Rubenstein, A. H., Recant, L.: Plasma proinsulin-like material in insulin treated diabetics. Horm. Metab. Res. 6, 439–443 (1974)
32. Logothetopoulos, J., Davidson, J. K., Haist, R. E., Best, C. H.: Degranulation of beta cells and loss of pancreatic insulin after infusions of insulin antibody or glucose. Diabetes 14, 493–500 (1965)
33. Gregor, W. H., Martin, J. M., Williamson, J. R., Lacy, P. E., Kipnis, D. M.: A study of the diabetic syndrome produced in rats by anti-insulin serum. Diabetes 12, 73–81 (1963)
34. Berson, S. A., Yalow, R. S.: Quantitative aspects of the reaction between insulin and insulin-binding antibody. J. Clin. Invest. 38, 1996–2016 (1959)
35. Phelps, R. L., Bergenstal, R., Freinkel, N., Rubenstein, A. H., Metzger, B. E., Mako, M.: Carbohydrate metabolism in pregnancy: XIII. Relationship between plasma insulin and proinsulin during late pregnancy in normal and diabetic subjects. J. Clin. Endocrinol. Metab. 41, 1085–1091 (1975)
36. Gorden, P., Hendricks, C. M., Roth, J.: Circulating proinsulin-like component in man: Increased proportion in hypoinsulinemic states. Diabetologia 10, 469–474 (1974)
37. Duckworth, W. C., Kitabchi, A. E.: The effect of age on plasma proinsulin-like material after oral glucose. J. Lab. Clin. Med. 88, 359–367 (1976)
38. Duckworth, W. C., Kitabchi, A. E.: Direct measurement of plasma proinsulin in normal and diabetic subjects. Am. J. Med. 53, 418–427 (1972)
39. Christiansen, Aa. H.: Radioimmunoelectrophoresis in the determination of insulin binding to IgG. Methodological studies. Horm. Metab. Res. 5, 147–154 (1973)

Received: November 1, 1976, and in revised form: April 26, 1977

Dr. Lise G. Heding
Novo Research Institute
Novo Alle
DK-2880 Bagsvaerd
Denmark

Reprinted from DIABETES
The Journal of The American Diabetes Association
Copyright 1978 by the American Diabetes Association, Inc.
Vol. 27, Supplement 1, Pg. 178-83

Insulin, C-peptide, and Proinsulin in Nondiabetics and Insulin-treated Diabetics

Characterization of the Proinsulin in Insulin-treated Diabetics

Lise G. Heding, Ph.D., Copenhagen

SUMMARY

Immunoreactive insulin (IRI), C-peptide, and proinsulin were determined in 46 fasting nondiabetic persons. The means ± S.D. for IRI, C-peptide, and proinsulin (in pmol/ml.) in the 46 nondiabetics were 0.054 ± 0.027, 0.39 ± 0.12, and 0.014 ± 0.009, respectively. The molar ratio between C-peptide and IRI varied from 4.0 to 16.7, with a mean value of 7.9 ± 3.4. Proinsulin constituted 5-36 per cent of IRI, with a mean of 20 ± 10 per cent.

Sera from 24 diabetics aged 17-67 years treated with conventional crystalline insulin for durations of from one to 29 years (mean: 15 years), all of whom had circulating insulin antibodies, were extracted with acid ethanol. Total IRI varied from 2.0 to 22.6 pmol/ml. C-peptide was detectable in only three patients (0.05, 0.08, and 0.15 pmol/ml.). In contrast, human proinsulin was found in all patients and ranged from 0.09 to 9.4 pmol/ml., corresponding to 1 to 45 per cent of total IRI. Species-specific immunoassays for bovine and porcine proinsulin revealed the presence of bovine proinsulin in 23 of the patients (mean 0.36 ± 0.20 pmol/ml.) and porcine proinsulin in eight patients (mean 0.15 ± 0.13 pmol/ml.). Total proinsulin constituted 5 to 64 per cent, mean 22 per cent, of total IRI, while human proinsulin amounted to 1 to 45 per cent, mean 10 per cent, of IRI. Although no insulin secretion could be detected in 21 of the patients, as judged by the amount of C-peptide, the B-cell was still able to secrete proinsulin. Insulin antibodies bind a portion of the secreted proinsulin, thus prolonging its half-life and increasing its concentration in serum. DIABETES 27 (Suppl. 1):178-83, 1978.

The normal human B-cell secretes, beside insulin, equimolar amounts of C-peptide and a few per cent of proinsulin,[187] but, for a number of reasons, the ratio between C-peptide, IRI, and proinsulin in peripheral blood of fasting nondiabetics has not yet been determined precisely in any large number of persons. Radioimmunoassays for insulin have been available since 1960, when Yalow and Berson[505] described the first assay. In recent years, several different radioimmunoassays have been described for human C-peptide[23,33,171,209,243,301] and proinsulin.[23,33,142,171,175,209,222,243,301] These assays can be used directly only in serum from persons without insulin

antibodies. However, many insulin-treated diabetics have circulating insulin antibodies as a result of previous treatment with conventional insulin. These antibodies bind a portion of the exogenous as well as the endogenous insulin, besides binding other molecules, such as proinsulin, which has an insulin moiety,[33,117,175,239] and this makes it necessary to extract the serum at low pH before analysis. This binding of human proinsulin to insulin antibodies results in its concentration's being higher than that of C-peptide [33,174,175,239] and invalidates a direct measurement of the latter. Furthermore, in the determination of proinsulin by gel filtration in serum extracts from patients treated with insulin, it has been shown that only up to 32 per cent of the proinsulin-like material was of human origin.[117]

The purpose of this study was dual: (1) To determine the ratio between immunoreactive insulin (IRI),

From the Novo Research Institute, Copenhagen, Denmark.

Address reprint requests to Lise G. Heding, Ph.D., Novo Research Institute, Novo Alle, DK-2880 Bagsvaerd, Copenhagen, Denmark.

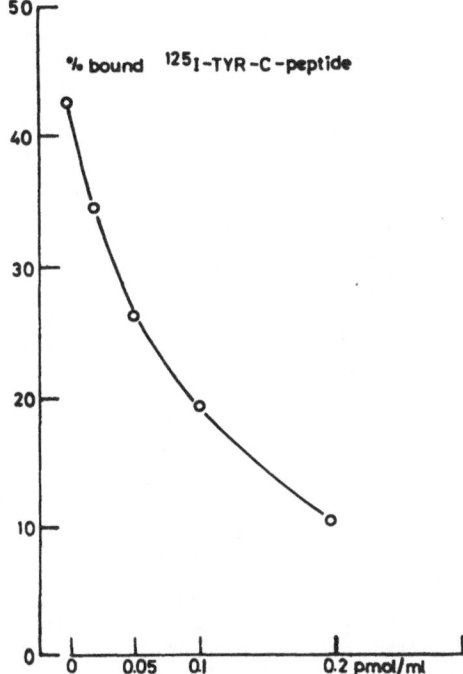

FIG. 1. Human proinsulin standard curve. The human proinsulin was bound to insulin antibodies coupled to Sepharose.

C-peptide, and proinsulin in fasting nondiabetics by highly sensitive and specific radioimmunoassays for the three peptides; and (2) To determine total IRI, C-peptide, and total proinsulin in long-term insulin-treated diabetics, as well as the fraction of total IRI comprised by proinsulin, and to characterize the composition of the proinsulin in respect to its species.

MATERIALS AND METHODS

Radioimmunoassays

Human C-peptide was determined after removal of proinsulin by Sepharose-bound insulin antibodies (S-AIS), by antiproinsulin serum M 1181, synthetic ^{125}I-Tyr-C-peptide, and synthetic C-peptide.[171] The detection limit was 0.03 pmol/ml. The S-AIS-bound human proinsulin was determined by use of human proinsulin (kindly donated by Prof. A. Rubenstein, Chicago) as the standard, ^{125}I-Tyr-C-peptide as the tracer, and antiserum M 1181 (see figure 1). Bovine and porcine proinsulins were determined on species-specific radioimmunoassays.[167] IRI in fasting non-diabetics was measured with anti-insulin guinea pig serum M 8170 (in a final dilution of 1:1,350,000), ^{125}I-porcine insulin (8 μU./ml.) as the tracer, and human insulin (2-25 μU./ml.) as the standard (see figure 2). The immunoassay procedure was that described earlier.[166] In extracts of diabetic sera, the usual immunoassay was used, with standards ranging from 5 to 100 μU./ml.[166]

Preparation and Extraction of Serum

Blood samples were taken after an overnight fast and, in the case of insulin-treated diabetics, before the morning injection. Serum was prepared and stored at $-18°$ C. Sera containing insulin antibodies were extracted at a pH of approximately 2 with acid ethanol.[166] The ethanol extract was poured into bottles and evaporated by blowing air at 25° C. The apparatus contained 100 samples at a time, and the evaporation was completed within two hours. The dry residue was dissolved in phosphate buffer containing albumin and used to measure the IRI, C-peptide, and human, bovine, and porcine proinsulins.[166]

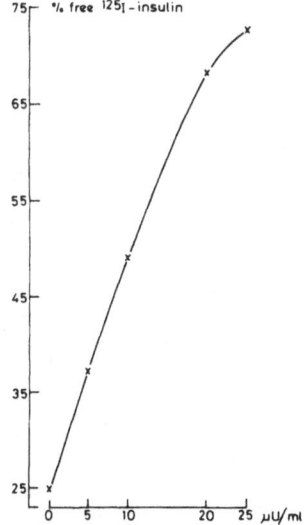

FIG. 2. Insulin standard curve. The anti-insulin guinea-pig serum was used in a final dilution of 1:1,350,000, the tracer in a final concentration of 8 μU./ml.

TABLE 1

IRI, C-peptide, and proinsulin in 46 fasting nondiabetics

	IRI		C-peptide	Proinsulin
	μU./ml.	pmol/ml.	pmol/ml.	pmol/ml.
Range	0-18	0-0.111	0.18-0.78	0-0.036
Mean	8.8	0.054	0.39	0.014
S.D.	4.3	0.027	0.12	0.009

Antibody Determinations

Insulin binding to IgG was determined by radioimmunoelectrophoresis[66] according to reference 17. Antibodies directed specifically against the C-peptide moiety of bovine and porcine proinsulin were determined after removal of all insulin-binding antibodies, as follows:

To 200 μl. of serum was added 600 μl. of a Sepharose-MC bovine insulin suspension[171,411] containing 1 mg. of Sepharose-bound insulin (S-I) per milliliter. After 20 hours' shaking at 4° C., the S-I-bound antibodies were removed by centrifugation and the supernatant used for proinsulin antibody determinations. The supernatant was checked for absence of insulin antibodies,[411] and in no case were residual amounts of these antibodies detected.

The determination of porcine and bovine proinsulin binding IgG was performed as follows: 50 μl. of supernatant was mixed with 50 μl. of barbital buffer and 50 μl. of either porcine or bovine ^{125}I-proinsulin (125 ng./ml., 25 μCi./μg.) in barbital buffer (ionic strength: 0.08, pH 8.6). The mixture was allowed to stand at 4° C. for 20 hours, then subjected to radioimmunoelectrophoresis in an agarose gel containing H-chain-specific antihuman IgG raised in rabbits, to measure the proinsulin bound to IgG.[66] The amount of proinsulin bound was calculated in nanograms per milliliter supernatant. Ten normal sera showed a mean nonspecific binding of 0.65 ng./ml., S.D. = 0.28 ng./ml. The results were corrected for this value of nonspecific binding, and only values over 1 ng./ml. ($\sim$ 3 S.D.) were considered positive.

Patients

The patient material comprised 24 insulin-treated diabetic outpatients of the Hvidøre Hospital (Head: Dr. C. Binder). There were 13 women and 11 men, aged 17-67 years (mean age: 45 $\pm$ 14 years). They had been treated for one to 29 years (mean: 15 $\pm$ 9 years) exclusively with the conventional four to five times crystallized insulins Lente, Rapitard, and Leotard. Their daily dose varied from 16 to 68 U. (mean: 37 $\pm$ 12 U.).

RESULTS

IRI, C-peptide, and Proinsulin in Nondiabetics

The detection limit (the smallest quantity of insulin that can be distinguished as significantly different from zero point = 2 S.D.) in the insulin assay shown in figure 2 was less than 1 μU./ml. The detection limit of the proinsulin assay (figure 1) was slightly below 0.01 pmol/ml., and that of the C-peptide assay was 0.03 pmol/ml.[171] Table 1 shows the results in the 46 nondiabetics. It appears that the molar concentration of C-peptide was much higher than that of insulin. Table 2 shows the calculated molar ratio between C-peptide and IRI as well as the per cent total IRI comprised by proinsulin. Human proinsulin showed approximately 66 per cent reactivity in the IRI assay—that is to say, 0.5 pmol/ml. of proinsulin reacted as well as 50 μU./ml. of insulin. In order to express the per cent total IRI comprised by proinsulin, the pmol/ml. estimates of proinsulin were converted into μU./ml. The molar ratio between C-peptide and IRI varied considerably, ranging from 4 to 16.7, and proinsulin comprised from 5 to 36 per cent IRI in this group.

Antibodies Against Insulin and the C-peptide Moiety of Bovine and Porcine Proinsulin in Insulin-treated Diabetics

Table 3 shows the capacity of insulin-binding IgG (mU./ml.), total extractable IRI, and the binding capacities of IgG directed against the C-peptide moiety of bovine and porcine proinsulin, respectively. All patients had insulin antibodies and, consequently, a highly elevated fasting total IRI. Twenty patients had IgG against porcine proinsulin and 19 had it against bovine proinsulin. There was no correlation between the proinsulin-binding IgG and the insulin-binding IgG, whereas a highly significant correlation could be established between the insulin-binding IgG and total IRI (P < 0.001).

Total IRI, C-peptide, and Bovine, Porcine, and Human Proinsulin in Insulin-treated Diabetics

Table 4 shows the concentrations of IRI, C-peptide, and the three proinsulins in the 24 pa-

TABLE 2

Molar ratio of C-peptide and IRI to proinsulin (μU./ml.) in per cent IRI (μU./ml.) in fasting nondiabetics

Number	C-peptide/IRI	Proinsulin in per cent IRI
of patients	43*	30†
Range	4.0-16.7	5-36
Mean	7.9	20
S.D.	3.4	10

*Three of the 46 persons had nondetectable IRI.

†Sixteen of the 46 persons had nondetectable IRI or proinsulin and were omitted.

TABLE 3

Total IRI, insulin-binding IgG, and IgG specifically directed against bovine and porcine proinsulins
in 24 long-term insulin-treated diabetics

| | Total IRI μU./ml. | Insulin mU./ml. | IgG binding capacity for | |
			Bovine proinsulin* ng./ml.	Porcine proinsulin* ng./ml.
Range	322-3,637	0.17-10.65	0.8-163	0.4-347
Median	751	2.99	15	26
Mean	1,054	3.67	33	55
S.D.	897	2.76	40	79

*These binding capacities were determined after complete removal of the insulin-binding IgG, which also binds proinsulins.

TABLE 4

Total IRI, C-peptide, and bovine, porcine, and human proinsulin in 24 fasting insulin-treated diabetics (all in pmol/ml.)

	Total IRI	C-peptide	Bovine proinsulin	Porcine proinsulin	Human proinsulin
Number of patients with significant concentration	24	3	23	8	24
Range	2.0-22.6	0-0.15	0.1-0.8	0-0.6	0.10-9.4
Mean	6.5	0.02	0.36	0.15	1.2
S.D.	5.6	0.03	0.20	0.13	2.2

tients. Only three of these patients had detectable C-peptide, while human proinsulin was found in all patients in concentrations far higher than the normal fasting values.

Table 5 shows that the total proinsulin (bovine + porcine + human) comprises as much as 64 per cent of the total IRI and that human proinsulin comprises 1 to 45 per cent of the total IRI. Bovine and porcine proinsulin showed the same reactivity as insulin in the IRI assay,[296] meaning that 1 pmol of proinsulin reacted approximately like 150 μU./ml. of insulin, and this was the factor used in converting the pmol/ml. measurements of exogenous proinsulin into μU./ml. for the calculation of proinsulin in per cent of total IRI.

DISCUSSION

It has been recognized that the ratio between IRI, C-peptide, and proinsulin found in peripheral blood is different from the ratio in which these three products are secreted from the B-cell. This has been shown with the greatest of clarity in the elegant study published by Horwitz et al.,[187] who demonstrated that C-peptide and insulin were secreted in equimolar concentrations, whereas proinsulin comprised approximately only 2.5 per cent of the IRI. It has been demonstrated that endogenous C-peptide has a half-time of 20 minutes as against only 10 minutes for IRI.[244] In another study, endogenous proinsulin was found to have a mean half-time of 17.2 minutes,

while the half-time of insulin was established at 4.8 minutes.[444] These findings indicate that the concentrations of C-peptide and proinsulin in the peripheral blood will comprise a higher percentage of the IRI than their concentrations on secretion from the B-cell. There is, however, considerable disagreement among investigators as to the ratio between C-peptide and IRI in fasting normal subjects. Thus, a mean molar ratio of 5.4 has been reported with a fasting C-peptide value of 0.6 pmol/ml.,[24] while another study reported a mean ratio of 24, with fasting C-peptide values varying between 1.3 and 2.8 pmol/ml.,[187] and yet another investigation found a mean ratio of 7.7 in 14 normals whose C-peptide was 0.37 pmol/ml.[172] In order to calculate a meaningful ratio, several prerequisites have to be met: (1) The IRI assay has to be very sensitive to ensure correct recordings of the low fasting values; and (2) The C-peptide assay should be specific and sensitive and not provide falsely high values because of the nonspecific reactivity of the antiserum with serum proteins.[103,171]

In the present study, we used an antiserum devoid

TABLE 5

Total proinsulin (sum of bovine, porcine, and human) and human proinsulin in per cent total IRI

| | Total IRI μU./ml. | Total proinsulin | | Human proinsulin | |
		μU./ml.	%	μU./ml.	%
Range	322-3,637	42-990	5-64	10-940	1-45
Mean	1,954	190	22	120	10
S.D.	897	223	15	219	14

of this nonspecific interference. There is considerable individual variation in the molar ratio of C-peptide to IRI, ranging from 4 to 16.7 (mean: 7.9). The few studies done so far on the half-time of C-peptide and insulin in man[244] provide no explanation for this high ratio. It may be that the conditions of the experiments were not imitative of the fasting state and that the liver degradation of the small amounts of insulin is proportionally higher in the fasting than in the stimulated state.

Proinsulin has been shown to constitute 2.5 per cent of the secreted IRI, a value probably somewhat underestimated since insulin was used as the standard,[1] while amounting to about 20 per cent in peripheral blood.[143,146,175,187] In the present study, proinsulin varied from 0 to 0.034 pmol/ml. in 46 fasting normals, with a mean value of 0 014 pmol/ml., constituting 5 to 36 per cent of the IRI (mean: 20 per cent).

In insulin-treated diabetics, the situation becomes far more complex because of the formation of antibodies as a consequence of treatment with conventional insulin,[27,411] and IRI, proinsulin, and C-peptide can be determined only after an acid extraction, if correct values for the three peptides are to be obtained.[121,175,239] It has been demonstrated that serum from patients with insulin antibodies may contain high amounts of human proinsulin-like material bound to antibodies,[33,117,175] and it has also been demonstrated that the presence of proinsulin interferes with the measurement of C-peptide because of the reactivity of the proinsulin in the C-peptide assay.[175,239] In a study in which C-peptide and proinsulin were measured in 73 juvenile insulin-treated diabetics aged five to 20 years, proinsulin was found in 42 per cent of the patients while C-peptide was detectable in 26 per cent, and the mean concentration of proinsulin was higher than that of C-peptide.[174] This clearly demonstrates that determination of C-peptide in diabetic sera containing insulin antibodies should be done only after its separation from proinsulin. In the present study, proinsulin up to a level of 9.4 pmol/ml. was found in all of 24 patients who had insulin antibodies, while C-peptide was detected in only three patients. Had C-peptide been determined directly, without removing proinsulin, a falsely high value of C-peptide would have conveyed the impression of the B-cell secreting both C-peptide and insulin.

On the other hand, it is obviously of interest to determine the amount of human proinsulin in insulin-treated diabetics as a source of supplementary information on B-cell function. In the group of patients described here, the concentration of human proinsulin ranged from 0.1 to 9.4 pmol/ml. and constituted, on the average, 10 per cent of total IRI. The high concentration of proinsulin is due to the fact that the insulin antibodies bind the proinsulin, prolong its half-life, and as a consequence thereof, increase its concentration to values much higher than they would have been in the absence of antibodies.[33] It is difficult, however, to explain the high per cent of total IRI constituted by human proinsulin. In the absence of C-peptide, most of the total IRI is of exogenous origin, and the high levels of human proinsulin could be due to an increased secretion of nonconverted proinsulin, unless insulin antibodies are far more avid in binding human proinsulin than insulin, as in the case of the two patients reported in reference 117. However, it appears that the measurement of proinsulin provides the means to follow B-cell secretory function even longer than the use of C-peptide assay. Diabetics treated with conventional insulin form insulin antibiodies and, because of the presence of bovine and porcine proinsulin in conventional insulin preparations, may also produce antibodies directed against the C-peptide moiety of the two proinsulins.[5,237,238] In the group of 24 patients treated for periods of one to 29 years, 20 had IgG directed against porcine proinsulin and 19 had IgG against bovine proinsulin. In order to determine these antibodies, it was of utmost importance to remove all the antibodies against insulin. This was ensured by adding a large excess of Sepharose-bound monocomponent bovine insulin and checking that the removal had been complete by a very sensitive method.[411] Figure 3 shows the many insulin and proinsulin components—both free and antibody-bound—that may be present in serum from insulin-treated patients. Total IRI comprises free and bound insulin of porcine, bovine, and human species and the three proinsulins, free as well as bound. A total-IRI measurement of maximum precision requires the following: a reproducible extraction proce-

$Ab_I \cdot I$	I
$Ab_I \cdot Pl_{bo}$	Pl_{bo}
$Ab_I \cdot Pl_{po}$	Pl_{po}
$Ab_I \cdot Pl_{hu}$	Pl_{hu}
$Ab_{Pl_{bo}} \cdot Pl_{bo}$	C
$Ab_{Pl_{po}} \cdot Pl_{po}$	

FIGURE 3

Insulin, proinsulin, and C-peptide components in insulin-treated diabetics with antibodies. Ab = antibodies, I = bovine, porcine, and human insulin, Pl = proinsulin, po = porcine, bo = bovine, hu = human, C = human C-peptide.

dure, an identical reactivity of human, bovine, and porcine insulin in the insulin immunoassay, and a high reactivity of proinsulin in the insulin immunoassay.

In the present study (1) the recovery in the extraction procedure was approximately 75 per cent, with an S.D. of about 3 per cent for triplicate extractions,[166] (2) one of the criteria for selection of the anti-insulin serum was that it should show identical reactions with the three species of insulin,[166] and (3) the bovine and porcine proinsulin showed identical (molar) reactions in the IRI assay,[296] while human proinsulin showed about 66 per cent reactivity.[175] The probable reason for these high degrees of reactivity is that a two-step assay was employed, and the standard or the sample was incubated with antiserum before addition of tracer. When gel filtration is used to separate proinsulin from insulin in acid ethanol extracts, total proinsulin will appear in the first IRI peak, and it was found that 10 to 62 per cent of the total proinsulin-like material was human proinsulin.[117] In the present study, the total proinsulin was characterized by specific radioimmunoassays for each species. It was found that human proinsulin was present in all patients, bovine proinsulin in 23, and porcine proinsulin in eight. Total proinsulin (μU./ml.) ranged from 42 to 990, comprising 5 to 64 per cent of the total IRI, which was in good agreement with the 10 to 62 per cent reported in reference 117. Human proinsulin in this study amounted to 10-940 μU./ml., or, 1 to 45 per cent of total IRI. When calculated in per cent of total proinsulin, the human proinsulin comprised 10 to 96 per cent. In comparison, values of 7 to 32 per cent were found in six patients.[112]

It can thus be concluded that determination of human proinsulin in acid extracts of diabetic sera containing insulin antibodies (obtained from patients treated with conventional insulin) cannot be performed by gel filtration followed by IRI determination of the various fractions because of the presence of exogenous proinsulin. Sera from patients treated with monocomponent insulin, which contains less than 1 ppm of proinsulin,[410] should present no such problem. The physiologic significance of the high amounts of circulating total proinsulin has yet to be established.

ACKNOWLEDGMENTS

The skillful technical assistance of Ms. Marianne Heyden, Ms. Lisbeth Petri Petersen, and Ms. Bente Hansen is gratefully acknowledged.

Reprinted from DIABETES
The Journal of The American Diabetes Association
Copyright 1978 by the American Diabetes Association, Inc.
Vol. 27, Supplement 1, Pg. 272-85

Reference Section
for C-peptide Symposium

[1]Ambler, R. P.: Enzymic hydrolysis with carboxypeptidases. Methods Enzymol. 11:155-66, 1967.

[2]Albano, J. D. M., Ekin, R. P., Maritz, G., and Turner, R. C.: A sensitive, precise radioimmunoassay of serum insulin relying on charcoal separation of bound and free hormone moieties. Acta Endocrinol. Copenhagen 70:487-509, 1972.

[3]Alberti, K. G. M. M., Record, C. O., Williamson, D. H., et al.: Metabolic changes in active chronic hepatitis. Clin. Sci. 42:591-605, 1972.

[4]Alberti, K. G. M. M., Hockaday, T. D. R., and Turner, R. C.: Small doses of intramuscular insulin in the treatment of diabetic coma. Lancet 2:515-22, 1973.

[5]Andersen, O. O.: Antibodies to proinsulin in diabetic patients treated with porcine insulin preparations. Acta Endocrinol. Copenhagen 73:304-13, 1973.

[8]Anderson, G. W., Zimmermann, J. E., and Callahan, F. M.: The use of esters of N-hydroxysuccinimide in peptide synthesis. J. Am. Chem. Soc. 86:1839-42, 1964.

[9]Anderson, G. W., Zimmermann, J. E., and Callahan, F. M.: Racemization control in the synthesis of peptides by the mixed carbonic-carboxylic anhydride method. J. Am. Chem. Soc. 88:1338-39, 1966.

[10]Archer, J. A., Gorden, P., and Roth, J.: Defect in insulin binding to receptors in obese man. J. Clin. Invest. 55:166-74, 1975.

[13]Bach, F. H., and van Rood, J. J.: The major histocompatibility complex—genetics and biology (third of three parts). New Engl. J. Med. 295:927-36, 1976.

[14]Baetens, D., Rufener, Cl., Srikant, C. B., Dobbs, R. E., Unger, R. H., and Orci, L.: Identification of glucagon-producing cells (A-cells) in dog gastric mucosa. J. Cell Biol. 69:455-64, 1976.

[15]Bagdade, J. D., Porte, D., Jr., and Bierman, E. L.: Hypertriglyceridemia. A metabolic consequence of chronic renal failure. New Engl. J. Med. 279:181-85, 1968.

[16]Bagdade, J. D., Bierman, E. L., and Porte, D.: Counterregulation of basal insulin secretion during alcohol hypoglycemia in diabetic and normal subjects. Diabetes 21:65-70, 1972.

[18]Bar, R. S., and Roth, J.: Insulin receptor status in diseased state of man. Arch. Intern. Med. 137:474-81, 1977.

[19]Barnes, A. J., Garbien, K., Crowley, M., and Bloom, A.: Effect of short and long term chlorpropamide on insulin release and blood glucose. Lancet 2:69-72, 1974.

[20]Barnes, A. J., Bloom, S. R., Alberti, K. G. M. M., Smythe, P., Alford, F. P., and Chisholm, D. J.: Ketoacidosis in pancreatectomized man. N. Engl. J. Med. 296:1250-53, 1977.

[21]Beischer, W., Raptis, S., Keller, L., Heinze, E., Schröder, K. E., and Pfeiffer, E. F.: Characterization of the residual beta-cell function in diabetics by a new C-peptide radioimmunoassay. 11th Annu. Meet. Eur. Assoc. Study Diabetes, Munich. Diabetologia 11:332, 1975. Abstract.

[22]Beischer, W., Kerner, W., Raptis, S., Keller, L., and Beischer, B.: Diabetes therapy and residual beta-cell function determined through C-peptide. Diabetologia 12:380, 1976. Abstract.

[23]Beischer, W., Keller, L., Maas, M., Schiefer, E., and Pfeiffer, E. F.: Human C-peptide, part I: radioimmunoassay. Klin. Wochenschr. 54:709-15, 1976.

[24]Beischer, W., Heinze, E., Keller, L., Raptis, S., Kerner, W., and Pfeiffer, E. F.: Human C-peptide, part II: clinical studies. Klin. Wochenschr. 54:717-25, 1976.

[25]Beischer, W., Raptis, S., Keller, L., Maas, M., Beischer, B., Feilen, K., and Pfeiffer, E. F.: Humanes C-peptid. Teil III: Sekretionsdynamik der Beta-Zellen erwachsener Diabetiker nach Glibenclamid-Glukose i.v. Klin. Woshenschr. Accepted for publication.

[26]Bennion, L. J., and Grundy, S. M.: Effects of diabetes mellitus on cholesterol metabolism in man. N. Engl. J. Med. 296:1365-71, 1977.

[27]Berson, S. A., Yalow, R. S., Bauman, A., Rothschild, M. A., and Newerly, K.: Insulin-I^{131} metabolism in human subjects: Demonstration of insulin binding globulin in the circulation of insulin treated subjects. J. Clin. Invest. 35:170-90, 1956.

[28]Berson, S. A., and Yalow, R. S.: Plasma insulin. In Diabetes Mellitus: Theory and Practice. Ellenberg, M., and Rifkin, H., Eds. New York, McGraw-Hill, 1970, pp. 308-67.

[29]Binder, C., Faber, O. K., and Lauritzen, T.: B-cell function and blood glucose in insulin dependent diabetes mellitus during the first years of treatment. Diabetologia 12:381, 1976.

[30]Binder, C., and Faber, O. K.: Residual beta-cell function and its metabolic consequences. Diabetes 27 (Suppl. 1):224-27, 1978.

[31]Blackard, W. G., and Nelson, N. C.: Portal and peripheral vein immunoreactive insulin concentrations before and after glucose infusion. Diabetes 19:302-06, 1970.

[32]Block, M. B., Mako, M., Steiner, D. F., and Rubenstein, A.

H.: Elevated circulating proinsulin levels in insulin requiring diabetics. J. Lab. Clin. Med. 44:31-32, 1971.

[33]Block, M. B., Mako, M. E., Steiner, D. F., and Rubenstein, A. H.: Circulating C-peptide immunoreactivity. Studies in normals and diabetic patients. Diabetes 21:1013-26, 1972.

[34]Block, M. B., Mako, M. E., Steiner, D. F., and Rubenstein, A. H.: Diabetic ketoacidosis. Evidence of C-peptide and proinsulin secretion following recovery. J. Clin. Endocrinol. Metab. 35:402-06, 1972.

[35]Block, M. B., Rosenfield, R. L., Mako, M. E., Steiner, D. F., and Rubenstein, A. H.: Sequential changes in beta-cell function in insulin-treated diabetic patients assessed by C-peptide immunoreactivity. N. Engl. J. Med. 288:1144-48, 1973.

[36]Block, M. B., Pildes, R. S., Mossabhoy, N. A., Steiner, D. F., and Rubenstein, A. H.: C-peptide immunoreactivity (CPR): a new method for studying infants of insulin-treated mothers. Pediatrics 53:923-28, 1974.

[37]Blundell, T., Dodson, G., Hodgkin, D., and Mercola, D.: Insulin: The structure in the crystals and its reflection in chemistry and biology. Adv. Protein Chem. 26:279-402, 1972.

[38]Blundell, T. L, Bedarkar, S., Rinderknecht, E., and Humbel, R. E.: Insulin-like growth factor: A model for the tertiary structure accounting for its immunoreactivity and receptor binding. Proc. Natl. Acad. Sci. USA: 1978, in press.

[39]Bornstein, J., and Lawrence, R. D.: Two types of diabetes mellitus, with and without available plasma insulin. Br. Med. J. 1:732, 1951.

[40]Boshell, B. R., Fox, O. J., and Roddam, R. F.: The effect of sulphonylurea agents on insulin secretion and insulin reserve. In Tolbutamide, After Ten Years. Butterfield, W. J. H., and von Westering, W., Eds. Amsterdam, Excerpta Medica, 1967, pp. 286-97.

[41]Bosnes, R. W., and Taussky, H. H.: On the colorimetric determinations of creatinine by the Jaffe reaction. J. Biol. Chem. 158:581-91, 1945.

[42]Bottazzo, G. F., Florin-Christensen, A., and Doniach, D.: Islet cell antibodies in diabetes mellitus with autoimmune polyendocrine deficiences. Lancet 2:1279-82, 1974.

[43]Bottazzo, G. F., Doniach, D., and Pouplard, A.: Humeral autoimmunity in diabetes mellitus. Acta Endocrinol. (Copenhagen), Suppl. 205:55-61, 1976.

[44]Bourdeau, J. E., and Carone, F. A.: Protein handling by the renal tubule. Nephron 13:22-34, 1974.

[45]Bridgen, J., Snary, D., Crumpton, M. J., Barnstable, C., Goodfellow, P., and Bodmer, W. F.: Isolation and N-terminal amino acid sequence of membrane bound human HLA-A and HL-B antigens. Nature 261:200-05, 1976.

[46]Bromer, W. W.: Proinsulin. BioScience 20:701, 1970.

[47]Brunzell, J. D., Robertson, R. P., Lerner, R. L., Hazzard, W. R., Ensinck, J., Bierman, E. G., and Porte, D.: Relationships between fasting plasma glucose levels and insulin secretion during intravenous glucose tolerance tests. J. Clin. Endocrinol. Metab. 42:222-29, 1976.

[*8]Brush, J. M.: Initial stabilization of the diabetic child. Am. J. Dis. Child. 67:429-44, 1944.

[50]Buschard, K., Christau, B., Christy, M., Nerup, J., Platz, P., Ryder, L. P., Svejgaard, A., Thomsen, M., and Bottazzo, G. F.: HLA, autoimmunity and juvenile diabetes mellitus. Diabetologia 12:382, 1976.

[51]Busse, W. D., Hansen, S. R., and Carpenter, F. H.: Carbonylbis (L-methionyl) insulin. A proinsulin analog which is convertible to insulin. J. Am. Chem. Soc. 96:5949-50, 1974.

[52]Cahill, G. F.: Physiology of insulin in man. The Banting Memorial Lecture 1971. Diabetes 20:785-99, 1971.

[53]Cahill, G. F., Etzwiler, D. D., and Freinkel, N.: Control and diabetes. N. Engl. J. Med. 294:1004-05, 1976.

[54]Camerini-Davalos, R., Root, H. F., and Marble, A.: Clinical experience with carbutamide (BZ 55). A progress report. Diabetes 6:74-77, 1957.

[55]Camu, F.: Hepatic balances of glucose and insulin in response to physiological increments of endogenous insulin during glucose infusions in dogs. Eur. J. Clin. Invest. 5:101-08, 1975.

[56]Canterbury, J. M., Bricker, L. A., Levey, G. S., Kozlovskis, P. L., Ruiz, E., Zull, J. E., and Reiss, E.: Metabolism of bovine parathyroid hormone. Immunological and biological characteristics of fragments generated by liver perfusion. J. Clin. Invest. 55:1245-53, 1975.

[58]Cecil, R. L.: On hypertrophy and regeneration of the islands of Langerhans. J. Exp. Med. 14:500-19, 1911.

[59]Chamberlain, M. J., and Stimmler, L.: The renal handling of insulin. J. Clin. Invest. 46:911-19, 1967.

[60]Chance, R. E., Ellis, R. M., and Bromer, W. W.: Porcine proinsulin: Characterization and amino acid sequence. Science 161:165-67, 1968.

[61]Chance, R. E.: Characterization of porcine proinsulin. Recent Progr. Horm. Res. 25:272-78, 1969.

[62]Chance, R. E.: Chemical, physical, biological and immunological studies on porcine proinsulin and related polypeptides. In Proceedings, 7th Congr. IDF, Buenos Aires, 1970. Rodriguez, R. R., and Vallance-Owen, J., Eds. Amsterdam, Excerpta Medica, 1971, pp. 292-305.

[63]Chance, R. E.: Amino acid sequences of proinsulins and intermediates. Diabetes 21 (Suppl. 2):461-67, 1972.

[64]Child, C. G., III: The Hepatic Circulation and Portal Hypertension. Philadelphia, Saunders, 1954.

[65]Chou, P. Y., and Fasman, G. D.: Prediction of protein conformation. Biochem. J. 13:222-45, 1974.

[66]Christiansen, A. H.: Radioimmunoelectrophoresis in the determination of insulin binding to IgG. Methodological studies. Horm. Metab. Res. 5:147-54, 1973.

[67]Christy, M., Nerup, J., Bottazzo, G. F., Doniach, D., Platz, P., Svejgaard, A., Ryder, L. P., and Thomsen, M.: Association between HLA-B8 and autoimmunity in juvenile diabetes mellitus. Lancet 2:142-43, 1976.

[68]Christy, M., Deckert, T., and Nerup, J.: Immunity and autoimmunity in diabetes mellitus. Clin. Endocrinol. Metab. 6:305-32, 1977,

[69]Chu, P. C., Conway, M. J., Krouse, H. A., and Goodner, C. J.: The pattern of response of plasma insulin and glucose to meals and fasting during chlorpropamide therapy. Ann. Intern. Med. 68:757-69, 1968.

[70]Clemens, A. H., Chang, P. H., and Myers, R. W.: Le développement d'un système automatique d'infusion d'insuline contrôlé par la glycémie, son système de dosage du glucose et ses algorithmes de contrôle. Journ. Annu. Diabetol. Hotel Dieu Paris, Flammarion, 1976, pp. 269-78.

[71]Collin, J., Taylor, R. M. R., and Johnston, I. D. A.: Carbohydrate tolerance with portal and systemic venous drainage of the pancreas. Br. J. Surg. 64:180-82, 1977.

[72]Collins, J. R., Lacy, W. W., Stiel, J. N., et al.: Glucose intolerance and insulin resistance in patients with liver disease. II. A study of etiological factors and evaluation of insulin actions. Arch. Intern. Med. 126:608-14, 1970.

[73]Conn, H. O., and Daughaday, W. H.: Cirrhosis and dia-

betes. V. Serum growth hormone levels in Laennec's cirrhosis. J. Lab. Clin. Med. 76:678-88, 1970.

[74]Constan, L., Mako, M., Juhn, D., and Rubenstein, A. H.: The excretion of proinsulin and insulin in urine. Diabetologia 11:119-23, 1975.

[75]Corrall, R. J. M., Thornley, P., Bhalla, I. P., and Duncan, L. J. P.: Observations on the efficacy and safety of glipizide, a new low dosage sulphonylurea. Acta Therapeut. 2:77-88, 1976.

[76]Couropmitree, C., Freinkel, N., Nagel, T. C., Horwitz, D. L., Metzger, B., Rubenstein, A. H., and Hahnel, R.: Plasma C-peptide and diagnosis of factitious hyperinsulinism: study of an insulin-dependent diabetic patient with "spontaneous" hypoglycemia. Ann. Intern. Med. 82:201-04, 1975.

[77]Craighead, J. E., and Nerup, J.: Etiology and pathogenesis of insulin-dependent diabetes mellitus. Workshop Report, Juvenile Diabetes Foundation. New York, 1977.

[78]Crapo, P. A., Reaven, G., and Olefsky, J.: Plasma glucose and insulin response to orally administered simple and complex carbohydrates. Diabetes 25:741-47, 1976.

[79]Cremer, G. M., Molnar, G. D., Taylor, W. F., Moxness, K. E., Service, F. J., Gatewood, L. C., Ackerman, R., and Rosevear, J. W.: Studies of diabetic instability. II. Tests of insulinogenic reserve with infusions of arginine, glucagon, epinephrine and saline. Metabolism 20:1083-98, 1971.

[81]Creutzfeldt, W., Frerichs, H., and Sickinger, K.: Liver diseases and diabetes mellitus. In Progress in Liver Diseases. Vol. 3. Popper, H., and Schnaffner, F., Eds. London, Heinemann, 1970, pp. 371-407.

[82]Creutzfeldt, W., Köbberling, J., and Neel, J. V.: The genetics of diabetes mellitus. Berlin, Heidelberg, New York, Springer-Verlag, 1976.

[83]Cudworth, A. G., and Woodrow, J. C.: Genetic susceptibility in diabetes mellitus: analysis of the H.L.A. association. Br. Med. J. 2:846-48, 1976.

[84]Cudworth, A. G., and Woodrow, J. C.: HL-A antigens and diabetes mellitus. Lancet 2:1153, 1974.

[85]Dausset, J., and Svejgaard, A. (eds.): HLA and Disease. Copenhagen, Munksgaard, 1977.

[86]Dausset, J., Hors, J., Contu, L., Schmid, M., Canivet, J., Cathelineau, G., Lestradet, H., and Baron, D.: Diabète insulinodépendant et HLA. Journ. Annu. Diabetol. Hotel Dieu. Flammarion, Médicine-Sciences 1977, pp. 78-94.

[87]Davies, B. J.: Disc electrophoresis. II. Method and application to human serum proteins. Ann. N.Y. Acad. Sci. 121:404-27, 1964.

[88]Dixon, K., Exon, P. D., and Malins, J. M.: Insulin antibodies and the control of diabetes. Q. J. Med. 44:543-53, 1975.

[89]Doar, J. W. H., Thompson, M. E., Wilde, C. E., and Sewell, P. F.: Influence of treatment with diet alone on oral glucose tolerance test and plasma sugar and insulin levels in patients with maturity onset diabetes mellitus. Lancet 1:1263-66, 1975.

[90]Dohan, F. C., and Lukens, F. D. W.: Experimental diabetes produced by the administration of glucose. Endocrinology 42:244-62, 1948.

[91]Doniach, I., and Morgan, A. G.: Islets of Langerhans in juvenile diabetes mellitus. Clin. Endocrinol. 2:233-48, 1973.

[92]Duckworth, W. C., Solomon, S. S., and Kitabchi, A. E.: Effect of chronic sulphonylurea therapy on plasma insulin and proinsulin levels. J. Clin. Endocrinol. Metab. 35:585-91, 1972.

[94]Enk, B., and Deckert, T.: Insulin secretion in insulin-requiring diabetics before and during insulin treatment. Abstract.

11th Annu. Meet. Eur. Assoc. Study Diabetes. Diabetologia 11:340, 1975.

[95]Faber, O. K., and Binder, C.: Personal communication.

[96]Faber, O. K., Binder, C., Markussen, J., Naithani, V. K., and Heding, L. G.: Plasma C-peptide in insulin requiring diabetes mellitus. Diurnal variation and response to oral glucose load during the first 9 months of treatment. Abstract. 11th Ann. Meet. Eur. Assoc. Study Diabetes. Munich. Diabetologia 11:340, 1975.

[97]Faber, O. K., Binder, C., Hendriksen, C., Drejer, J., and Heding, L. G.: Plasma C-peptide response to glucagon as a measure of β-cell function in insulin dependent diabetes mellitus. Diabetes 25 (Suppl. 1):329, 1976. Abstract.

[98]Faber, O. K., Binder, C., Lauritzen, T., and Heding, L. G.: Preserved B-cell function and blood glucose control in insulin dependent diabetes mellitus. Diabetes 25 (Suppl. 1):362, 1976. Abstract.

[99]Faber, O. K., Markussen, J., Naithani, V. K., and Binder, C.: Production of antisera to synthetic benzyloxycarbonyl-C-peptide of human proinsulin. Hoppe-Seylers Z. Physiol. Chem. 357:751-57, 1976.

[100]Faber, O. K., and Binder, C.: B-cell function and blood glucose control in insulin dependent diabetics within the first month of insulin treatment. Diabetologia 13:263-68, 1977.

[101]Faber, O. K., and Binder, C.: C-peptide response to glucagon. A test for the residual β-cell function in diabetes mellitus. Diabetes 26:605-10, 1977.

[102]Faber, O. K., and Binder, C.: Plasma C-peptide during the first year of insulin dependent diabetes mellitus. Proc. IXth IDF Congr., New Delhi, 1976. Amsterdam, Excerpta Medica, 1977, in press.

[103]Faber, O. K., Binder, C., Markussen, J., Heding, L. G., Naithani, V. K., Kuzuya, H., Blix, P., Horwitz, D., and Rubenstein, A. H.: Characterization of seven C-peptide antisera. Diabetes 27 (Suppl. 1):207-09, 1978.

[104]Faber, O. K., Kehlet, H., Madsbad, S., and Binder, C.: Kinetics of human C-peptide in man. Diabetes 27 (Suppl. 1): 207-09, 1978.

[105]Fajans, S. S., Floyd, J. C., Pek, S., Knopf, R. F., and Jacobsen, M.: Effect of protein meals on plasma insulin in mildly diabetic patients. Diabetes 18:523-28, 1969.

[106]Fajans, S. S., Floyd, J. C., Taylor, C. E., and Pek, S.: Heterogeneity of insulin responses in latent diabetes. Trans. Assoc. Am. Physicians 87:83-94, 1975.

[107]Feldman, J. M., and Lebovitz, H. E.: Appraisal of the extrapancreatic actions of sulfonylureas. Arch. Intern. Med. 123:314-22, 1969.

[108]Feldman, J. M., and Lebovitz, H. E.: Endocrine and metabolic effects of glibenclamide: evidence for an extrapancreatic mechanism of action. Diabetes 20:745-55, 1971.

[109]Felig, P., Brown, W., Levine, R. A., et al.: Glucose homeostasis in viral hepatitis. N. Engl. J. Med. 283:1436-40, 1970.

[110]Felig, P., and Wahren, J.: Influence of endogenous insulin secretion of splanchnic glucose and amino acid metabolism in man. J. Clin. Invest. 50:1702-11, 1971.

[113]Felig, P., Wahren, J., Sherwin, R., and Palaiologos, G.: Amino acid and protein metabolism in diabetes mellitus. Arch. Intern. Med. 137:507-13, 1977.

[114]Field, J. B., Webster, M., and Drapanas, T.: Evaluation of factors regulating hepatic control of insulin homeostasis. Proc. 60th Annu. Meet. Am. Soc. Clin. Invest. 47:33a-34a, 1968.

[115]Field, J. B.: Extraction of insulin by liver. Annu. Rev. Med. 24:309-14, 1973.

[116]Fine, J.: Glucose content of normal urine. Br. Med. J. 1:1209-14, 1965.

[117]Fink, G., Cresto, J. C., Gutman, R. A., Lavine, R. L., Rubenstein, A. H., and Recant, L.: Plasma proinsulin-like material in insulin treated diabetics. Horm. Metab. Res. 6:439-43, 1974.

[118]Finkelstein, S., Zeller, E., and Walford, R.: No relation between HL-A and juvenile diabetes. Tissue Antigens 2:74-77, 1972.

[119]Floyd, J. C., Fajans, S. S., Conn, J. W., Thiffault, C., Knopp, R. F., and Guntsche, E.: Secretion of insulin induced by amino acids and glucose in diabetes mellitus. J. Clin. Endocrinol. Metab. 28:266-76, 1968.

[120]Floyd, J. C., Jr., and Fajans, S. S.: A newly recognized human pancreatic islet polypeptide: concentrations in healthy subjects and in patients with diabetes mellitus. Diabetes 25 (Suppl. 1):330, 1976. Abstract.

[121]Franckson, J. R. M., and Ooms, H. A.: The glomerular clearance of exogenous insulin. Horm. Metab. Res. 5:75-79, 1973.

[122]Frank, B. H., and Veros, A. J.: Physical studies on proinsulin-association behaviour and conformation in solution. Biochem. Biophys. Res. Commun. 32:155-60, 1968.

[123]Frank, B. H., Veros, A. J., and Pekar, A. H.: Physical studies on proinsulin. A comparison of the titration behaviour of the tyrosine residues in insulin and proinsulin. Biochem. J. 11:4926-31, 1972.

[124]Freychet, P., Roth, J., and Neville, D. M.: Monoiodoinsulin: demonstration of its biological activity and binding to fat cells and liver membranes. Biochem. Biophys. Res. Commun. 43:400-08, 1971.

[125]Frier, B. M., Faber, O. K., Binder, C., and Elliott, H. L.: The effect of residual insulin secretion on exocrine pancreatic function in juvenile-onset diabetes. Diabetes 26 (Suppl. 1):369, 1977. Abstract.

[126]Fujita, Y., Herron, A. L., and Seltzer, H. S.: Confirmation of impaired insulin response to glycemic stimulus in nonobese mild diabetics. Diabetes 24:17-27, 1975.

[127]Fullerton, W. W., Potter, R., and Low, B. W.: Proinsulin: Crystallization and preliminary x-ray diffraction studies. Proc. Natl. Acad. Sci. U.S.A. 66:1213-19, 1970.

[128]Følling, I., and Norman, N.: Hyperglycemia, hypoglycemic attacks and production of anti-insulin antibodies without previous immunization: immunological and functional studies in a patient. Diabetes 21:814-26, 1972.

[129]Gabbay, K. H., Deluca, K., Fisher, J. N., Mako, M. E., and Rubenstein, A. H.: Familial hyperproinsulinemia: An autosomal dominant defect. N. Engl. J. Med. 294:911-15, 1976.

[130]Gabbay, K. H., Wolff, J., Bergenstal, R., Mako, M., and Rubenstein, A. H.: Familial hyperproinsulinemia: Partial characterization of abnormal proinsulin. Diabetes 26 (Suppl. 1):376, 1977. Abstract.

[131]Gainer, H., Pengloh, V., and Sarne, Y.: Biosynthesis of neuronal peptides. In Peptides in Neurobiology. Gainer, H., Ed. New York, Plenum Press, 1977, pp. 185-219.

[132]Gamble, D. R., Kinsley, M. L., Fitzgerald, M. G., Bolton, R., and Taylor, K. W.: Viral antibodies in diabetes mellitus. Br. Med. J. 3:627-30, 1969.

[133]Geiger, R., Jäger, G., and König, W.: Zur Synthese von Peptiden mit Eigenschaften des Human-Proinsulin-C-peptids.

VI. Darstellung der Gesamtsequenz des Human-Proinsulin-C-peptids und seines (Glu⁹, Gln¹¹)-Analogen. Chem. Ber. 106:2347-52, 1973.

[134]Genuth, S. M.: Plasma insulin and glucose profiles in normal, obese and diabetic persons. Ann. Intern. Med. 79:812-22, 1973.

[135]Gepts, W.: Pathologic anatomy of the pancreas in juvenile diabetes mellitus. Diabetes 14:619-33, 1965.

[136]Gepts, W.: Pathology of islet tissue in human diabetes. In Handbook of Physiology. Vol. 1. Endocrine Pancreas. Steiner, D. F., and Freinkel, N., Eds. Baltimore, Williams and Wilkins, 1972, pp. 289-303.

[137]Gepts, W.: The endocrine pancreas. Functional morphology and histology. In Insulin. Islet Pathology, Islet Function, Insulin Treatment. Luft, R., Ed. Acta Med. Scand. Suppl. 601:9-52, 1976.

[138]Gepts, W., De Mey, J., and Marichal-Pipeleers, M.: Hyperplasia of "pancreatic polypeptide" cells in the pancreas of juvenile diabetics. Diabetologia 13:27-34, 1977.

[139]Goldman, J., Baldwin, D., Rubenstein, A. H., Klink, D. D., Blackard, W. G., Fisher, L. K., Roe, T. F., and Schnure, J. J.: The syndrome of hypoglycemia and autoimmune insulin antibodies. Submitted for publication.

[141]Gonen, B., Rochman, H., Tanega, S. P., and Rubenstein, A. H.: Correlation between HbAI levels and clinical assessment of diabetic control. Diabetes 26 (Suppl. 1):368, 1977. Abstract.

[142]Gorden, P., and Roth, J.: Plasma insulin: Fluctuations in the "big" insulin component in man after glucose and other stimuli. J. Clin. Invest. 48:2225-34, 1969.

[143]Gorden, P., Sherman, B., and Roth, J.: Proinsulin-like component of circulating insulin in the basal state and in patients and hamsters with islet cell tumors. J. Clin. Invest. 50:2113-22, 1971.

[145]Gorden, P., Hendricks, C. M., and Roth, J.: Circulating proinsulin-like component in man: increased proportion in hypoinsulinemic states. Diabetologia 10:469-74, 1974.

[146]Gorden, P., Roth, J., Hendricks, C. M., and Kahn, C. R.: Plasma proinsulin-like components. Isr. J. Med. Sci. 10:1212-21, 1974.

[147]Graf, R., and Porte, D., Jr.: Glycosylated hemoglobin as an index of glycemia independent of plasma insulin in normal and diabetic subjects. Diabetes 26 (Suppl. 1):368, 1977 (Abstr.).

[148]Grajwer, L. A., Pildes, R. S., Horwitz, D. L., and Rubenstein, A. H.: Control of juvenile diabetes mellitus and its relationship to endogenous insulin secretion as measured by C-peptide immunoreactivity. J. Pediatr. 90:42-48, 1977.

[149]Greco, A. V., Fedeli, G., Ghirlanda, G., et al.: Behaviour of pancreatic glucagon, insulin and HGH in liver cirrhosis, after arginine and I.V. glucose. Acta Diabetol. Lat. 11:330-39, 1974.

[150]Greenwood, R. H., Mahler, R. F., and Hales, C. N.: Improvement in insulin secretion in diabetes after diazoxide. Lancet 1:444-47, 1976.

[151]Grodsky, G., and Forsham, P.: An immunochemical assay of total extractable insulin in man. J. Clin. Invest. 39:1070-79, 1960.

[152]Grodsky, G. M., Feldman, R., Toreson, W. E., and Lee, J. C.: Diabetes mellitus in rabbits immunized with insulin. Diabetes 15:579-85, 1966.

[153]Guillemin, R., Vargo, T., Rossier, J., Minick, S., Ling, N., Rivier, C., Vale, W., and Bloom, F.: β-endorphin and adrenocorticotropin are secreted concomitantly by the pituitary gland. Science 197:1367-69, 1977.

[155]Guttmann, S.: Synthèse du glutathion et de l'oxytocine à l'aide d'un nouveau groupe protecteur de la fonction thiol. Helv. Chim. Acta 49:83-96, 1966.

[156]Gutman, R. A., Lazarus, N. R., Penhos, J. C., et al.: Circulating proinsulin-like material in patients with functioning insulinomas. N. Engl. J. Med. 284:1003-08, 1971.

[157]Gutman, R. A., Lazarus, N. R., and Recant, L.: Electrophoretic characterization of circulating human proinsulin and insulin. Diabetologia 8:136-40, 1972.

[159]Hadden, D. R., Montgomery, D. A. D., Skelly, R. J., Trimble, E. R., Weaver, J. A., Wilson, E. A., and Buchanan, K. D.: Maturity onset diabetes mellitus: response to intensive dietary management. Br. Med. J. 3:276-78, 1975.

[160]Hagen, C., Faber, O. K., Binder, C., and Alberti, K. G. M. M.: Lack of metabolic effect of C-peptide in normal subjects and juvenile diabetic patients. Acta Endocrinol. (Copenhagen) 85 (Suppl. 209):25, 1977.

[161]Hahn, H. J., Menzel, R., Gottschling, H. D., and Jahr, D.: Enhancement of glucose-stimulated insulin secretion from isolated rat pancreatic islets by human insulin antibodies. Acta Endocrinol. Copenhagen 83:123-32, 1976.

[162]Hansen, A. P., and Johansen, K.: Diurnal patterns of blood glucose, serum free fatty acids, insulin, glucagon and growth hormone in normals and juvenile diabetics. Diabetologia 6:27-33, 1970.

[163]Hansen, B. W., Koerker, D. J., Goodner, C. J., Brown, A. C., and Rubenstein, A. H.: Evidence for oscillating secretion of insulin in M. mulatta, its metabolic consequences and damping with starvation. Endocrinology 100 (Suppl.):104, 1977.

[164]Hanwi, G. J.: Oral hypoglycaemic agents. Curr. Med. Digest 33:1184-92, 1966.

[165]Hartroft, W. S., and Wrenshall, G. A.: Correlation of beta cell granulation with extractable insulin of the pancreas. Studies in adult human diabetics and non diabetics. Diabetes 4:1-7, 1955.

[166]Heding, L. G.: Determination of total serum insulin (IRI) in insulin-treated diabetic patients. Diabetologia 8:260-66, 1972.

[167]Heding, L. G., Larsen, U. D., Markussen, J., Jørgensen, K. H., and Hallund, O.: Radioimmunoassays for human, pork and ox C-peptides and related substances. Horm. Metab. Res. 5 (Suppl.):40-44, 1974.

[168]Heding, L. G., Turner, R. C., and Harris, E.: C-peptide, proinsulin and insulin responses to fish-insulin induced hypoglycemia in the diagnosis of insulinomas. Diabetes 24 (Suppl. 2):412, 1975.

[169]Heding, L. G., and Munkgaard Rasmussen, S.: Insulin, C-peptide, and proinsulin in nondiabetics and insulin-treated diabetics. Diabetes 27 (Suppl. 1):173-83, 1978.

[170]Heding, L. G., Dahl Larsen, U., Markussen, J., and Naithani, V. K.: Human plasma C-peptide assay, sources of error and examples of its use. 11th Annu. Meet. Eur. Assoc. Study Diabetes, Munich. Diabetologia 11:348, 1975. Abstract.

[171]Heding, L. G.: Radioimmunological determination of human C-peptide in serum. Diabetologia 11:541-48, 1975.

[172]Heding, L. G., and Rasmussen, S. M.: Human C-peptide in normal and diabetic subjects. Diabetologia 11:201-06, 1975.

[173]Heding, L. G.: Human bovine and porcine proinsulin in insulin-treated diabetics. Diabetologia 13:400, 1977. Abstract.

[174]Heding, L. G., and Ludvigsson, J.: Human proinsulin in insulin-treated juvenile diabetics. Acta Paediatr. Scand. In press.

[175]Heding, L. G.: Specific and direct radioimmunoassay for human proinsulin in serum. Diabetologia 13:467-74, 1977.

[176]Hedner, L. P., and Nordén, A.: Analysis of variance applied to blood glucose values for assessing insulin preparations and state of lability in diabetes. Diabetologia 5:108-15, 1969.

[177]Hendriksen, C., Faber, O. K., Drejer, J., and Binder, C.: Prevalence of residual B-cell function in insulin-treated diabetics evaluated by the plasma C-peptide response to intravenous glucagon. Diabetologia 13:280-88, 1977.

[178]Hernandez, A., Zorilla, E., and Gershberg, H.: Decreased insulin production, elevated growth hormone levels, and glucose intolerance in liver disease. J. Lab. Clin. Med. 73:25-33, 1969.

[179]Hershko, A., Mamont, P., Shields, R., and Tompkins, G. M.: Pleiotypic response. Nature New Biol. London 232:206-11, 1971.

[180]Hetenyi, G., Norwich, K. H., and Zelin, S.: Analysis of the glucoregulatory system in dogs. Am. J. Physiol. 224:635-42, 1973.

[181]Hirata, Y., and Ishizu, H.: Elevated insulin-binding capacity of serum proteins in a case of spontaneous hypoglycemia and mild diabetes not treated with insulin. Tohoku J. Exp. Med. 107:277-86, 1977.

[182]Holdsworth, C. D., Nye, L., and King, E.: The effect of portacaval anastomosis on oral carbohydrate tolerance and on plasma insulin levels. Gut 13:58-63, 1972.

[183]Holman, R. R., and Turner, R. C.: Diabetes: the quest for basal normoglycaemia. Lancet 1:469-74, 1977.

[184]Holman, R. R., and Turner, R. C.: Basal normoglycaemia achieved with chlorpropamide in mild diabetes. Submitted for publication.

[185]Honzl, J., and Rudinger, J.: Amino acids and peptides. XXXIII. Nitrosyl chloride and butyl nitrite as reagents in peptide synthesis by the azide method: suppression of amide formation. Collect. Czech. Chem. Commun. 26:2333, 1961.

[186]Horwitz, D. L., Starr, J. I., Rubenstein, A. H., and Steiner, D. F.: Serum connecting peptide: an indicator of beta cell secretory function. Diabetes 22 (Suppl. 1):298, 1973. Abstract.

[187]Horwitz, D. L., Starr, J. I., Mako, M. E., Blackard, W. G., and Rubenstein, A. H.: Proinsulin, insulin and C-peptide concentrations in human portal and peripheral blood. J. Clin. Invest. 55:1278-83, 1975.

[188]Horwitz, D. L., Rubenstein, A. H., Reynolds, C., Molnar, G. D., and Yanaihara, N.: Prolonged suppression of insulin release by insulin induced hypoglycemia: demonstration by C-peptide assay. Horm. Metab. Res. 7:449-52, 1975.

[189]Horwitz, D. L., Kuzuya, H., and Rubenstein, A. H.: Circulating serum C-peptide: a brief review of diagnostic implications. N. Engl. J. Med. 295:207-09, 1976.

[190]Horwitz, D. L., Rubenstein, A. H., and Katz, A. I.: Quantitation of human pancreatic beta-cell function by immunoassay of C-peptide in urine. Diabetes 26:30-35, 1977.

[191]Hunter, W. M., and Greenwood, F. C.: Preparation of iodine[131]labelled growth hormone of high specific activity. Nature London 194:495-96, 1962.

[192]Ichihara, K., Shima, K., Saito, Y., Nonaka, K., Tarui, S., and Nishikawa, M.: Mechanism of hypoglycemia observed in a patient with insulin autoimmune syndrome. Diabetes 26:500-06, 1977.

[193]Ikeda, Y., Ando, N., Minami, N., and Ide, Y.: B-cell function of insulin-dependent young onset diabetics assessed by C-peptide immunoreactivity. Diabetologia 11:351-52, 1975.

[194]Ingelfinger, F. J.: Debates on diabetes. N. Engl. J. Med. 296:1228-30, 1977.

[195]Irvine, W. J., Clarke, B. F., Scarth, L., Cullen, D. R., and Duncan, L. J. P.:Thyroid and gastric autoimmunity in patients with diabetes mellitus. Lancet 2:163-68, 1970.

[196]Irvine, W. J., Gray, R. S., and McCallum, C. J.: Pancreatic islet cell antibody as a marker for asymptomatic and latent diabetes and prediabetes. Lancet 2:1098-1102, 1976.

[197]Irvine, W. J., McCallum, C. J., Gray, R. S., and Duncan, L. J. P.: Clinical and pathogenic significance of pancreatic-islet-cell antibodies in diabetics treated with oral hypoglycaemic agents. Lancet 1:1025-27, 1977.

[197a]Irvine, W. J., McCallum, C. J., Gray, R. S., Campbell, C. J., Duncan, L. J. P., Farquhar, J. W., Vaughan, H., and Morris, P. J.: Pancreatic islet-cell antibodies in diabetes mellitus correlated with the duration and type of diabetes, coexistent autoimmune disease and HLA type. Diabetes 26:138-47, 1977.

[198]Ito, S., and Kobayashi, S.: Immunohistochemical demonstration of glucagon- and GLI-containing cells in the canine gut and pancreas. Arch. Histol. Jap. 39:193-202, 1976.

[199]Izzo, J. L., Crump, S. L., and Kunz, W.: A clinical comparison of modified insulins. J. Clin. Invest. 29:1514-27, 1950.

[200]Jackson, R. L.: Juvenile diabetes. In Diabetes. Proc. 8th Congr. IDF. Malaisse, W. J., and Pirart, J., Eds. Amsterdam, Excerpta Medica, 1974, pp. 520-31.

[201]Jahnke, K., Miss, H. D., and Drost, H.: Kriterien und Bewertung der Diabeteseinstellung. Dtsch. Med. Wochenschr. 99:870-83, 1974.

[202]James, R. C., and Chase, G. R.: Evaluation of some commonly used semi-quantitative methods for urinary glucose and ketone determinations. Diabetes 23:474-79, 1974.

[203]Jersild, M.: Insulin zinc suspension: Four years' experience. Lancet 2:1009-13, 1956.

[204]Johnston, D. G., and Alberti, K. G. M. M.: Carbohydrate metabolism in liver disease. In Clinics in Endocrinology and Metabolism, Disorders of Carbohydrate Metabolism excluding Diabetes. Alberti, K. G. M. M., Ed. London, Philadelphia, Toronto. Saunders, 1976, pp. 657-702.

[205]Johnston, D. G., Alberti, K. G. M. M., Faber, O. K., Binder, C., and Wright, R.: Hyperinsulinism of hepatic cirrhosis: diminished degradation or hypersecretion? Lancet 1:10-13, 1977.

[206]Jones, R. H., Dion, D. I., Ellis, M. J., Sönksen, P. H., and Brandenburg, D.: Biological properties of chemically modified insulin. I. Biological activity of proinsulin and insulin modified at A1-glycine and B29-lysine. Diabetologia 12:601-08, 1976.

[207]Kaden, M., Harding, P., and Field, J. B.: Effect of intraduodenal glucose administration on hepatic extraction of insulin in the anesthetized dog. J. Clin. Invest. 52:2016-28, 1973.

[208]Kanazawa, Y., Kuzuya, T., Ide, T., and Kosaka, K.: Plasma insulin responses to glucose in femoral, hepatic, and pancreatic veins in dogs. Am. J. Physiol. 211:442-48, 1966.

[209]Kaneko, T., Oka, H., Munemura, M., Oda, T., Yamashita, K., Suzuki, S., Yanaihara, N., Hashimoto, T., and Yanaihara, J.: Radioimmunoassay of human proinsulin C-peptide using synthetic human connecting peptide. Endocrinol. Jap. 21:141-45, 1974.

[210]Kaneko, T., Munemura, M., Oka, H., Oda, T., Suzuki, H., Yasuda, H., Yanaihara, N., Nakagawa, S., and Makebe, K.: Demonstration of C-peptide immunoreactivity in various body fluids and clinical evaluation of the determination of urinary C-peptide immunoreactivity. Endocrinol. Jap. 22:207-12, 1975.

[211]Karakash, C., Assimacopoulos-Jeannet, F., and Jeanrenaud, B.: An anomaly of insulin removal in perfused livers of obese-hyperglycemic (ob/ob) mice. J. Clin. Invest. 57:1117-24, 1976.

[212]Karam, J. A., Grodsky, G. M., and Forsham, P. H.: Excessive insulin response to glucose in obese subjects as measured by immunochemical assay. Diabetes 12:197-204, 1963.

[213]Karam, J. H., Matin, S. B., and Forsham, P. H.: Antidiabetic drugs after the University Group Diabetes Program (UGDP). Annu. Rev. Pharmacol. 15:351-66, 1975.

[214]Katz, A. I., and Rubenstein, A. H.: Metabolism of proinsulin, insulin, and C-peptide in the rat. J. Clin. Invest. 52:1113-21, 1973.

[215]Kaufmann, R. L., Assal, J. Ph., Soeldner, J. S., Wilmshurst, E. G., Lemaire, J. R., Gleason, R. E., and White, P.: Plasma lipid levels in diabetic children. Effect of diet restricted in cholesterol and saturated fats. Diabetes 24:672-79, 1975.

[216]Kehlet, H., and Brandt, M.: C-peptide and insulin in plasma during blockade of the hyperglycemic response to surgery by epidural analgesia. Diabetologia 12:402, 1976. Abstract.

[217]Kemmler, W., Steiner, D. F., and Borg, J.: Studies on the conversion of proinsulin to insulin. III. Studies in vitro with a crude secretion granule fraction isolated from islets of Langerhans. J. Biol. Chem. 248:4544-51, 1973.

[218]Kijinuma, H., Kuzuya, T., and Ide, T.: Effects of hypoglycemic sulfonamides on glucagon and insulin secretion in ducks and dogs. Diabetes 23:412-17, 1974.

[220]Kipnis, D. M.: Nutrient regulation of insulin secretion in human subjects. Diabetes 21:606-16, 1972.

[221]Kitabchi, A. E.: The biological and immunological properties of pork and beef insulin, proinsulin, and connecting peptides. J. Clin. Invest. 49:979-87, 1970.

[222]Kitabchi, A. E., Duckworth, W. C., Brush, J. S., and Heinemann, M.: Direct measurement of proinsulin in human plasma by the use of an insulin-degrading enzyme. J. Clin. Invest. 50:1792-99, 1971.

[223]Kitabchi, A. E., Duckworth, W. C., Stentz, F. B., and Yu, S. S.: Properties of proinsulin and related polypeptides. In CRC critical reviews in biochemistry. Harte, R., Ed. Cleveland, Chemical Rubber Company, 1972, pp. 59-94.

[224]Kitabchi, A. E.: Proinsulin and C-peptide: A review. Metabolism 26:547-87, 1977.

[225]Klachko, D. M., and Burns, T. W.: Observations on glucose homeostasis using continuous monitoring. Horm. Metab. Res. 7 (Suppl.):64-71, 1977.

[226]Klimt, C. R. et al.: Standardization of the oral glucose tolerance test. Report of the Committee on Statistics of the American Diabetes Association. Diabetes 18:299-307, 1969.

[227]Ko, A. S. C., Smyth, D. G., Markussen, J., and Sundby, F.: The amino acid sequence of the C-peptide of human proinsulin. Eur. J. Biochem. 20:190-99, 1971.

[228]Kock, N. G., Roding, B., Hahnloser, P., Tibblin, S., and Worthington, G. S., Jr.: The effect of glucagon on hepatic blood flow. An experimental study in the dog. Arch. Surg. 100:147-49, 1970.

[229]Koenig, R. J., Peterson, C. M., Kilo, C., Cerami, A., and Williamson, J. R.: Hemoglobin A1c as an indicator of the degree of glucose intolerance in diabetes. Diabetes 25:230-32, 1976.

[230]Kopetz, K., and Wehrmann, P.: Stoffwechselverän-

derungen nach Glucose- und Tolbutamidbelastung bei Lebercirrhosen und ihre Rolle in der Pathogenese des hepatogenen Diabetes Mellitus. Klin. Wochenschr. 48:1265-76, 1970.

[231] Krarup, N., and Larsen, J. A.: The effect of glucagon on hepatosplanchnic hemodynamics, functional capacity, and metabolism of the liver in cats. Acta Physiol. Scand. 91:42-52, 1974.

[232] Kudo, M., Toyota, T., Abe, K., Kimura, K., and Goto, Y.: Effects of synthetic rat C-peptide 1 and 2 on insulin and glucagon secretion from the isolated perfused rat pancreas. In Current Topics in Diabetes Research, Abstracts, IX Cong. IDF. Bajaj, J. S., Ed. Amsterdam, Excerpta Medica, 1976, p. 16.

[233] Kühl, C.: Glucose metabolism during and after pregnancy in normal and gestational women. I. Influence of normal pregnancy on serum glucose and insulin concentration during basal fasting conditions and after a challenge with glucose. Acta Endocrinol. Copenhagen 79:709-19, 1975.

[234] Kühl, C., Andersen, O., Jensen, S. L., and Nielsen, O. V.: Effect of ethanol on the glucose-mediated insulin release in triply catheterized anesthetized pigs. Diabetes 25:752-57, 1976.

[235] Kühl, C.: Serum proinsulin in normal and gestational diabetic pregnancy. Diabetologia 12:295-300, 1976.

[236] Kühl, C., Faber, O. K., Hornnes, P., and Jensen, S. L.: C-peptide metabolism and the liver. Diabetes 27 (Suppl. 1): 197-200, 1978.

[237] Kumar, D., and Miller, L. V.: Proinsulin-specific antibodies in human sera. Diabetes 22:361-66, 1973.

[238] Kumar, D., and Miller, L. V.: The prevalence of proinsulin-specific antibodies in diabetic patients. Horm. Metab. Res. 5:1-3, 1973.

[239] Kuzuya, H., Blix, P. M., Horwitz, D. L., Steiner, D. F., and Rubenstein, A. H.: Determination of free and total insulin and C-peptide in insulin-treated diabetics. Diabetes 26:22-29, 1977.

[240] Kuzuya, H., Blix, P. M., Horwitz, D. L., Rubenstein, A. H., Steiner, D. F., Binder, C., and Faber, O. K.: Heterogeneity of circulating C-peptide. J. Clin. Endocrinol. Metab. 44:952-62, 1977.

[241] Kuzuya, H., Chance, R., Steiner, D. F., and Rubenstein, A. H.: On the preparation and characterization of the standard materials for natural human proinsulin and C-peptide. Diabetes 27 (Suppl. 1):161-69, 1978.

[242] Kuzuya, H., Blix, P. M., Horwitz, D. L., Rubenstein, A. H., and Steiner, D. F.: Heterogeneity of circulating C-peptide: Human C-peptide radioimmunoassays with three different antisera. Diabetes 27 (Suppl. 1):184-91, 1978.

[243] Kuzuya, T., Matsuda, A., Saito, T., and Yoshida, S.: Human C-peptide immunoreactivity (CPR) in blood and urine—evaluation of a radioimmunoassay method and its clinical applications. Diabetologia 12:511-18, 1976.

[244] Kuzuya, T., and Matsuda, A.: Disappearance rate of endogenous human C-peptide from blood. Diabetologia 12:519-21, 1976.

[245] Larsson, L. I., Holst, J., Håkanson, R., and Sundler, F.: Distribution and properties of glucagon immunoreactivity in the digestive tract of various mammals: an immunohistochemical and immunochemical study. Histochemistry 44:281-90, 1975.

[246] Larsson, Y., Häger, A., and Ludvigsson, J.: Insulin antibodies and therapeutic control in juvenile diabetics after transfer to monocomponent insulin (MCI). Acta Endocrinol. (Copenhagen) 85 (Suppl. 209):41, 1977. Abstract.

[248] Layne, E.: Protein estimation with the Folin-ciocalten reagent. Methods Enzymol. 3:448-50, 1957.

[249] Lazarus, N. R., Tanese, T., Gutman, R., and Recant, L.: Synthesis and release of proinsulin and insulin by human insulinomatissue. J. Clin. Endocrinol. 30:273-81, 1970.

[250] Lendrum, R., Walker, G., and Gamble, D. R.: Islet-cell antibodies in juvenile diabetes mellitus of recent onset. Lancet 1:880-83, 1975.

[251] Lendrum, R., Nelson, P. G., Pyke, D. A., Walker, G., and Gamble, D. R.: Islet cell, thyroid and gastric autoantibodies in diabetic identical twins. Br. Med. J. 1:553-55, 1976.

[252] Lendrum, R., Walker, G., Cudworth, A. G., Theophanides, C., Pyke, D. A., Bloom, A., and Gamble, D. R.: Islet cell antibodies in diabetes mellitus. Lancet 2:1273-76, 1976.

[253] Lendrum, R., Walker, G., Cudworth, A. G., Woodrow, J. C., and Gamble, D. R.: HLA-linked genes and islet cell antibodies in diabetes mellitus. Br. Med. J. 1:1565-67, 1976.

[254] Lerner, R. L., and Porte, D.: Epinephrine: selective inhibition of the acute insulin response to glucose. J. Clin. Invest. 50:2453-57, 1971.

[255] Lerner, R. L., and Porte, D.: Acute and steady srate insulin response to glucose in nonobese diabetic subjects. J. Clin. Invest. 51:1624-31, 1972.

[256] Lev-Ran, A., and Goldman, J. A.: Brittle diabetes in pregnancy. Diabetes 26:926-30, 1977.

[257] Lev-Ran, A.: Clinical observations on brittle diabetes. Arch. Intern. Med., in press.

[258] Lewis, S. B., Wallin, J. D., Kuzuya, H., Murray, W. K., Coustan, D. R., Daane, T. A., and Rubenstein, A. H.: Circadian variation of serum glucose, C-peptide immunoreactivity and free insulin in normal and insulin-treated diabetic pregnant subjects. Diabetologia 12:343-50, 1976.

[259] Lickley, H. L. A., Chisholm, D. J., Rabinovitch, A., et al.: Effects of portacaval anastomosis on glucose tolerance in the dog: evidence of an interaction between the gut and the liver in oral glucose disposal. Metabolism 24:1157-68, 1975.

[260] Liljenquist, J. E., Chiasson, J.-L., Jennings, A. S., Horwitz, D. L., and Rubenstein, A. H.: Insulin suppression of insulin secretion in man. Demonstration by C-peptide assay. Diabetes 24 (Suppl. 2):403, 1975.

[261] Lindberg, B., and Darle, N.: Effect of glucagon on hepatic circulation in the pig. Arch. Surg. 111:1379-83, 1976.

[262] Ludvigsson, J., and Heding, L. G.: C-peptide in children with juvenile diabetes. A preliminary report. Diabetologia 12:627-30, 1976.

[263] Ludvigsson, J.: Metabolic control in juvenile diabetes mellitus. The influence of some clinical, biochemical and sociopsychological factors. Linköping University, medical dissertation 42, 1976.

[264] Ludvigsson, J., and Heding, L. G.: C-peptide in diabetic children after stimulation with glucagon compared with fasting C-peptide levels in non-diabetic children. Acta Endocrinol. (Copenhagen) 85:364-71, 1977.

[265] Ludvigsson, J., Heding, L. G., Larsson, Y., and Leander, E.: C-peptide in juvenile diabetics beyond the postinitial remission period. Relation to clinical manifestations at onset of diabetes, remission and diabetic control. Acta Paediatr. Scand. 66:177-84, 1977.

[266] Ludvigsson, J., Säfwenberg, J., and Heding, L. G.: HLA-types: C-peptide and insulin antibodies in juvenile diabetes. Diabetologia 13:13-17, 1977.

[268]Ludvigsson, J., and Heding, L. G.: Proinsulin as a measure of beta cell function in children with juvenile diabetes. Abstr., 13th Annu. Meet. Eur. Assoc. Study Diabetes. Diabetologia 13:415, 1977.

[270]Ludvigsson, J.: Socio-psychological factors and metabolic control in juvenile diabetes. Acta Paediatr. Scand. In press.

[272]MacCallum, W. G.: Hypertrophy of the islands of Langerhans in diabetes mellitus. Am. J. Med. Sci. 133:432-40, 1907.

[273]McCarthy, S. T., Harris, E., and Turner, R. C.: Glucose control of basal insulin secretion in diabetes. Diabetologia 13:93-97, 1977.

[274]MacCuish, A. C., Barnes, E. W., Irvine, W. J., and Duncan, L. J. P.: Antibodies to pancreatic islet cells in insulin-dependent diabetics with coexistent autoimmune disease. Lancet 2:1529-31, 1974.

[275]McGarry, J. D., and Foster, D. W.: Hormonal control of ketogenesis. Biochemical considerations. Arch. Intern. Med. 137:495-501, 1977.

[276]McIlroy, M., Walsh, C. H., Doyle, C., et al.: Glucose tolerance in viral hepatitis. Ir. J. Med. Sci. 145:3-6, 1976.

[278]Maclean, N., and Ogilvie, R. F.: Observations on the pancreatic islets of young diabetic subjects. Diabetes 8:83-91, 1959.

[279]McNair, P., Madsbad, S., Christiansen, C., Transbøl, I., Faber, O. K., and Binder, C.: Bone mineral loss in diabetes mellitus. 12th Annu. Meet. Scand. Soc. Study Diabetes Uppsala 1977. Acta Endocrinol. (Copenhagen) 85 (Suppl. 209):45, 1977.

[280]Madison, L. L., and Kaplan, N.: The hepatic binding of I-131-labeled insulin in human subjects during a single trans-hepatic circulation. J. Lab. Clin. Med. 52:927, 1958.

[281]Madison, L. L., Combes, B., Unger, R. H., et al.: The relationship between the mechanism of action of the sulphonylureas and the secretion of insulin into the portal circulation. Ann. N.Y. Acad. Sci., 74:548-56, 1959.

[282]Madsbad, S., Faber, O. K., Binder, C., and Alberti, K. G. M. M.: Endogenous insulin secretion and the metabolism of β-hydroxybutyrate in insulin dependent diabetes mellitus. 12th Annu. Meet. Scand. Soc. Study Diabetes Uppsala 1977. Acta Endocrinol. (Copenhagen) 85 (Suppl. 209):1732, 1977.

[283]Madsbad, S., McNair, P., Christiansen, P., Faber, O., Transbøl, I., and Binder, C.: Prevalence of residual β-cell function in insulin-dependent diabetics in relation to age at onset and duration of diabetes. Diabetes 27 (Suppl. 1):260-62, 1978.

[284]Mains, R. E., Eipper, B. A., and Ling, N.: Common precursor to corticotrophins and endorphins. Proc. Natl. Acad. Sci. U.S.A. 74:3014-18, 1977.

[285]Mako, M., Block, M., Starr, J., et al.: Proinsulin in chronic renal and hepatic failure: a reflection of the relative contribution of the liver and kidney to its metabolism. Clin. Res. 21:631, 1973.

[286]Marble, A.: Criteria of control. In Diabetes Mellitus: Diagnosis and Treatment. New York, American Diabetes Association, 1964, pp. 69-71.

[287]Marco, J., Diego, J., Villanueva, M. L., et al.: Elevated plasma glucagon levels in cirrhosis of the liver. N. Engl. J. Med. 289:1107-11, 1973.

[288]Markussen, J.: Structural changes involved in the folding of proinsulin. Int. J. Peptide Protein Res. 3:201-07, 1971.

[289]Markussen, J., Sundby, F., Smyth, D. G., and Ko, A.: Preparation of human C-peptide. Horm. Metab. Res. 3:229-32, 1971.

[290]Markussen, J., and Sundby, F.: Rat-proinsulin C-peptides,

amino acid sequences. Eur. J. Biochem. 25:153-62, 1972.

[291]Markussen, J., and Schiff, H. E.: Molecular parameters of C-peptide from bovine proinsulin. Int. J. Peptide Protein Res. 5:69-72, 1973.

[292]Markussen, J., and Sundby, F.: Isolation and amino acid sequence of the C-peptide duck proinsulin. Eur. J. Biochem. 34:401-08, 1973.

[293]Markussen, J., and Heding, L.: Reduction and reoxidation of bovine proinsulin, effect of pH, zinc ions, temperature and concentration. Int. J. Peptide Protein Res. 6:245-52, 1974.

[294]Markussen, J., and Vølund, A.: A new method of calculating evolutionary rates of proteins applied to insulin and C-peptides. Int. J. Peptide Protein Res. 6:79-86, 1974.

[295]Markussen, J., and Vølund, A.: Conformational analysis of circular dichroism spectra of insulin, proinsulin and C-peptides by non-linear regression. Int. J. Peptide Protein Res. 7:47-56, 1975.

[296]Markussen, J., and Heding, L. G.: Separation of the two double-chain bovine intermediates of the proinsulin-insulin conversion. I. Chemical, immunochemical, circular dichroism and biological characterization. Int. J. Peptide Protein Res. 8:597-606, 1976.

[297]Marliss, E. B., Murray, F. T., Stokes, E. F., Zinman, B., Nakhooda, A. F., Denoga, A., Leibel, B. C., and Albisser, A. M.: Normalization of glycemia in diabetics during meals with insulin and glucagon delivery by the artificial pancreas. Diabetes 26:663-72, 1977.

[298]Martin, J. M., and Lacey, P. E.: The prediabetic period in partially pancreatectomized rats. Diabetes 12:238-42, 1963.

[299]Megyesi, C., Samols, E., and Marks, V.: Glucose tolerance and diabetes in chronic liver disease. Lancet 2:1051-56, 1967.

[301]Melani, F., Rubenstein, A. H., Oyer, P. E., and Steiner, D. F.: Identification of proinsulin and C-peptide in human serum by a specific immunoassay. Proc. Natl. Acad. Sci. 67:148-55, 1970.

[302]Melani, F., Ryan, W. G., Rubenstein, A. H., and Steiner, D. F.: Proinsulin secretion by a pancreatic beta-cell adenoma: proinsulin and C-peptide secretion. N. Engl. J. Med. 283:713-19, 1970.

[303]Merimee, T. J., and Tyson, J. E.: Stabilization of plasma glucose during fasting: normal variations in two separate studies. N. Engl. J. Med. 291:1275-78, 1974.

[304]Mirouze, J., Bernard, R., Sany, C., Satingher, A.: L'hyperglycémie alimentaire dans le diabète sucré (étude par enregistrement continu). Pathol.-Biol. (Paris) 13:152-62, 1965.

[305]Mirouze, J., Collard, F., Selam, J. L., and Pham, T. C.: Continuous blood glucose monitoring in insulin-treated diabetes. Blood glucose monitoring. Horm. Metab. Res. Suppl. 7:77-86, 1977.

[306]Misbin, R. I., Merimee, T. J., and Lowenstein, J. M.: Insulin removal by isolated perfused rat liver. Am. J. Physiol. 230:171-77, 1976.

[307]Miyamoto, S., Koizumi, J., Ota, M., Uamada, S., Inoue, T., Ueno, T., Mibayashi, Y., Mabuchi, H., and Takeda, R.: Clinical studies on the serum C-peptide responses to oral administration of 50 g glucose. J. Jap. Diab. Soc. 19:22-29, 1976.

[308]Mogensen, C. E.: Renal function changes in diabetes. Diabetes 25 (Suppl. 2):872-79, 1976.

[309]Molnar, G. D., Ackerman, E., Rosevear, J. W., Gatewood, L. C., and Moxness, K. E.: Continuous blood glucose analysis in ambulatory fed subjects. I. General methodology. Mayo Clin. Proc. 43:833-51, 1968.

[310]Molnar, G. D., Taylor, W. F., and Ho, M. M.: Day-to-day variation of continuously monitored glycemia: a further measure of diabetic instability. Diabetologia 8:342-48, 1972.

[311]Molnar, G. D., and Read, R. C.: Hypoglycemia and body temperature. J.A.M.A. 227:916-21, 1974.

[312]Molnar, G. D., Taylor, W. F., and Langworthy, A.: On measuring the adequacy of diabetes regulation: comparison of continuously monitored blood glucose patterns with values at selected time points. Diabetologia 10:139-43, 1974.

[313]Molnar, G. D.: Unstable diabetes: concepts of its nature and treatment based on continuous blood glucose monitoring studies. IDF Int. Congr. Bruxelles. Excerpta Medica. Series No. 312, 546-59, 1974.

[314]Molnar, G. D., and Reynolds, C.: Diurnal glucose variability and hormonal regulation. Blood glucose monitoring. Horm. Metab. Res. Suppl. 7:148-57, 1977.

[315]Molnar, G. D.: Metabolic derangements of diabetes—the challenge of normalization. IDF Int. Congr. New Delhi. Excerpta Medica Series 1977. In press.

[316]Mondon, C. E., Olefsky, J. M., Dolkas, C. B., and Reaven, G. M.: Removal of insulin by perfused rat liver: effect of concentration. Metabolism 24:153-60, 1975.

[317]Morgan, C. R., and Lazarow, A.: Immunoassay of insulin: two antibody system. Plasma insulin levels of normal, subdiabetic and diabetic rats. Diabetes 12:115-26, 1963.

[318]Morris, P. J., Vaughan, H., Irvine, W. J., McCallum, F. J., Gray, R. S., Campbell, C. J., Duncan, L. J. P., and Farquhar, J. W.: HLA and pancreatic islet cell antibodies in diabetes. Lancet 2:652-53, 1976.

[319]Mortimore, G. E., Tietze, F., and Stetten, D., Jr.: Metabolism of insulin-I[131]. Studies in isolated, perfused rat liver and hind-limb preparations. Diabetes 8:307-14, 1959.

[320]Mortimer, G. E., and Tietze, F.: Studies on the mechanism of capture and degradation of insulin-I[131] by the cyclically perfused rat liver. Ann. N.Y. Acad. Sci. 82:329-37, 1959.

[321]Moxness, K. E., Molnar, G. D., Taylor, W. F., Owen, C. A., Jr., Ackerman, E., and Rosevear, J. W.: Studies of diabetic instability. I. Immunoassay of human insulin in plasma containing antibodies to pork and beef insulin. Metabolism 20:1074-82, 1971.

[322]Munoz Barragan, L., Rufener, C., Srikant, C. B., Dobbs, R. E., Shannon, W. A., Jr., Baetens, D., and Unger, R. H.: Immunocytochemical evidence for glucagon-containing cells in the human stomach. Horm. Metab. Res. 9:37-39, 1977.

[323]Murthy, D. Y. N., Guthrie, R. A., Womack, W. N., and Jackson, R. L.: Chemical and early overt diabetes mellitus in children. I. Effect of glucagon on "insulin reserve." Diabetes 18:679-85, 1969.

[325]Naithani, V. K.: Immunoassays of the nine synthetic porcine C-peptide sequences. Horm. Metab. Res. 5:53, 1973.

[326]Naithani, V. K.: Studies on polypeptides. IV. The synthesis of C-peptide of human proinsulin. Hoppe-Seyler's Z. Physiol. Chem. 354:659-72, 1973.

[327]Naithani, V. K., Dechesne, M., Markussen, J., and Heding, L. G.: Studies on polypeptides V. Improved synthesis of human proinsulin C-peptide and its benzyloxycarbonyl derivative. Circular dichroism and immunological studies of human C-peptide. Hoppe-Seyler's Z. Physiol. Chem. 356:997-1010, 1975.

[328]Naithani, V. K., Dechesne, M., Markussen, J., Heding, L. G., and Larsen, U. D.: Studies on polypeptides. VI. Synthesis,

circular dichroism and immunological studies of tyrosol C-peptide of human proinsulin. Hoppe-Seyler's Z. Physiol. Chem. 356:1305-12, 1975.

[329]Neijadlik, D. C., Dube, A. H., and Adamko, S. M.: Glucose measurements and clinical correlations. J.A.M.A. 224:1734-36, 1973.

[330]Nerup, J., Andersen, O. O., Bendixen, G., Egeberg, J., and Poulsen, J. E.: Antipancreatic cellular hypersensitivity in diabetes mellitus. Diabetes 20:424-27, 1971.

[331]Nerup, J., Platz, P., Andersen, O. O., Christy, M., Lyngsøe, J., Poulsen, J. E., Ryder, L. P., Thomsen, M., Staub Nielsen, L., and Svejgaard, A.: HL-A antigens and diabetes mellitus. Lancet 2:864-66, 1974.

[332]Nerup, J., Cathelineau, C., Seignalet, J., and Thomsen, M.: HLA endocrine diseases. In HLA and disease. Dauset, J., and Svejgaard, A., Eds. Copenhagen, Munksgaard, 1977, pp. 149-67.

[333]Nicol, D. S. H. W., and Smith, L. F.: Amino acid sequence of human insulin. Nature 181:483-85, 1960.

[334]Nieschlag, E., Kremner, G. J., and Mussgnug, U.: Insulin, Glucosetoleranz und freie Fettsaüre während und nach akuter Hepatitis. Klin. Wschr. 48:381-83, 1970.

[335]Nikkilä, E. A., Huttunen, J. K., and Ehnholm, C.: Post-heparin plasma lipoprotein lipase and hepatic lipase in diabetes mellitus. Relationship to plasma triglyceride metabolism. Diabetes 26:11-21, 1977.

[336]Nishiitsutsuji-Uwo, J. M., Ross, B. D., and Krebs, H. A.: Metabolic activities of the isolated perfused rat kidney. Biochem. J. 103:852-62, 1967.

[337]Nolan, C., Margoliash, E., Peterson, J. D., and Steiner, D. F.: The structure of bovine proinsulin. J. Biol. Chem. 246:2780-95, 1971.

[338]Oakley, N. W., Harrigan, P., Kissbah, A. H., Kissin, E. A., and Adams, P. W.: Factors affecting insulin response to glucagon in man. Metabolism 21:1001-07, 1972.

[339]Olefsky, J. M., Farquhar, J. W., and Reaven, G. M.: Relationship between fasting plasma insulin level and resistance to insulin-mediated glucose uptake in normal and diabetic subjects. Diabetes 22:507-13, 1973.

[340]Ooms, H. A.: Emploi des Insulines Radioiodées comme Traceurs en Biologie. Bruxelles, Arscia Editions, 1973.

[341]Orci, L., Baetens, D., Rufener, Cl., Amberdt, M., Ravazzola, M., Studer, P., Malaisse-Lagae, F., and Unger, R. H.: Hypertrophy and hyperplasia of somatostatin-containing D-cells in diabetes. Proc. Natl. Acad. Sci. U.S.A. 73:1338-42, 1976.

[342]Oyama, H., Horino, M., Matsumura, S., Kobayshi, K., and Suetsugu, N.: Immunological half-life of porcine proinsulin C-peptide. Horm. Metab. Res. 7:520-21, 1975.

[343]Oyer, P. E., Cho, S., Peterson, J. D., and Steiner, D. F.: Studies on human proinsulin. Isolation and amino acid sequence of human pancreatic C-peptide. J. Biol. Chem. 246:1375-86, 1971.

[344]Palumbo, P. J., Molnar, G. D., Taylor, W. F., Moxness, K. E., and Tauxe, W. N.: Insulin antibody binding in diabetes mellitus and factitious hypoglycemia. Mayo Clin. Proc. 44:725-37, 1969.

[347]Parving, H.-H., Noer, I., Deckert, T., et al.: The effect of metabolic regulation on microvascular permeability to small and large molecules in short-term diabetes. Diabetologia 12:161-66, 1976.

[348]Patzelt, C., Chan, S. J., Duguid, J., Hortin, G., Keim, P.,

Heinrikson, R. L., and Steiner, D. F.: Biosynthesis of polypeptide hormones in intact and cell-free system. FEBS Proc. 1977, in press.

[349] Paulsen, E. P., and Koury, M.: Hemoglobin A1c levels in insulin-dependent and -independent diabetes mellitus. Diabetes 25 (Suppl. 2):890-96, 1976.

[351] Pekar, A. H., and Frank, B. H.: Conformation of proinsulin. A comparison of insulin and proinsulin self-association at neutral pH. Biochemistry 11:4013-16, 1972.

[352] Perley, M. J., and Kipnis, D. M.: Plasma insulin responses to oral and intravenous glucose: studies in normal and diabetic subjects. J. Clin. Invest. 46:1954-62, 1967.

[353] Peterson, C. M., Koenig, R. J., Jones, R. L., Saudek, C. D., and Cerami, A.: Correlation of serum triglyceride levels and hemoglobin A1c concentration in diabetes mellitus. Diabetes 26:507-09, 1977.

[354] Peterson, J. D., Nehrlich, S., Oyer, P. E., and Steiner, D. F.: Determination of the amino acid sequence of the monkey, sheep, and dog proinsulin C-peptides by a semi-micro Edman degradation procedure. J. Biol. Chem. 247:4866-71, 1972.

[355] Peterson, P. A., Evrin, P. E., and Berggard, I.: Differentiation of glomerular, tubular and normal proteinuria. Determinations of urinary excretion of β_2-macroglobulin, albumin, and total protein. J. Clin. Invest. 48:1189-98, 1969.

[356] Pfeiffer, E. F., Schöffling, K., Steigerwald, H., Ditschuneit, H., and Heubel, F.: Die Bedeutung der einmaligen Tablettenbelastung für die Indikationsstellung der oralen Diabetestherapie. Deutsch. Med. Wochenschr. 82:1544, 1957.

[357] Pfeiffer, E. F., Pfeiffer, M., Ditschuneit, H., and Chang-Su-Ahn: Über die Bestimmung von Insulin im Blute am epididymalen Fettanhang der Ratte mit Hilfe markierter Glukose. II. Experimentelle und klinische Erfahrungen. Klin. Wochenschr. 37:1239-45, 1959.

[358] Pfeiffer, E. F., Ditschuneit, H., and Ziegler, R.: Über die Bestimmung von Insulin im Blute am epididymalen Fettanhang der Ratte mit Hilfe markierter Glukose. IV. Die Dynamik der Insulinsekretion des Stoffwechselgesunden und des Altersdiabetikers nach wiederholten Belastung mit Glukose, Sulfonylharnstoffen und menschlichem Wachstumshormon, ein Beitrag zur Pathogenese des menschlichen Alterdiabetes. Klin. Wochenschr. 39:415-26, 1961.

[359] Pfeiffer, E. F., and Raptis, S.: Controlled extenstion of oral antidiabetic therapy on former insulin dependent diabetics by means of the combined i.v. glibenclamide-glucose-response-test. Diabetologia 8:41-47, 1972.

[360] Pfeiffer, E. F., Raptis, S., and Schröder, K. E.: Einmalige intravenöse Glibenclamid-Glukose-Belastung als Vorhersagetest. Deutsch. Med. Wochenschr. 99:1281-94, 1974.

[361] Pfeiffer, E. F., Beischer, W., and Kerner, W.: The artificial endocrine pancreas in clinical research. Horm. Metab. Res. Suppl. 7:95-114, 1977.

[363] Porte, D., and Pupo, A. A.: Insulin responses to glucose: evidence for a two pool system in man. J. Clin. Invest. 48:2309-19, 1969.

[364] Porte, D., Jr., and Bagdade, J. D.: Human insulin secretion: An integrated approach. Ann. Rev. Med. 21:219-40, 1970.

[365] Porte, D., Bagdade, J. D., and Lerner, R.: Basal insulin secretion: the abnormality of obesity and its relation to a two-compartmental model for insulin release. IDF, Buenos Aires. Excerpta Medica Int. Cong. Series. 544-50, 1971.

[366] Portugal-Alvarez, J. de, Clamagirand, C. P., Souto, J. M.,

et al.: Relationship between growth hormone, insulin and urinary oestrogens in hepatic cirrhosis. Acta Diabetol. Lat. 11:1-8, 1974.

[370] Rabkin, R., and Colwell, J. A.: The renal uptake and excretion of insulin in the dog. J. Lab. Clin. Med. 73:893-900, 1969.

[371] Rabkin, R., Simon, N. M., Steiner, S., and Colwell, J. A.: Effect of renal disease on renal uptake and excretion of insulin in man. N. Engl. J. Med. 282:182-87, 1970.

[372] Rabkin, R., Rubenstein, A. H., and Colwell, J. A.: Glomerular filtration and proximal tubular absorption of insulin 125I. Am. J. Physiol. 223:1093-96, 1972.

[373] Rabkin, R., Kitabchi, A., Young, J., and Barker, J.: The renal handling of insulin. Kidney Int. 10:573, 1976. Abstract.

[374] Rayfield, E. J., Pulini, M., Golub, A., Rubenstein, A. H., and Horwitz, D. L.: Nonautonomous function of a pancreatic insulinoma. J. Clin. Endocrinol. Metab. 43:1307-11, 1976.

[375] Reaven, G., and Dray, J.: Effect of chlorpropamide on serum glucose and immunoreactive insulin concentrations in patients with maturity-onset diabetes mellitus. Diabetes 16:487-92, 1967.

[376] Reaven, G. M., Bernstein, R., Davis, B., and Olefsky, J. M.: Non ketotic diabetes mellitus: insulin deficiency or insulin resistance? Am. J. Med. 60:80-88, 1976.

[377] Record, C. O., Alberti, K. G. M. M., Williamson, D. H., et al.: Glucose tolerance and metabolic changes in human viral hepatitis. Clin. Sci. Molec. Med. 45:677-90, 1973.

[378] Record, C. O., Chase, R. A., Alberti, K. G. M. M., et al.: Disturbances in glucose metabolism in patients with liver damage due to paracetamol overdose. Clin. Sci. Molec. Med. 49:473-79, 1975.

[379] Regeur, L., and Binder, C.: The correlation between plasma C-peptide and kidney function. 12th Annu. Meet. Eur. Assoc. Study Diabetes Helsinki. Diabetologia 12:416, 1976 (Abstr.).

[380] Reynolds, C., Molnar, G. D., Horwitz, D. L., Rubenstein, A. H., Taylor, W., and Jian, N.-S.: Abnormalities of endogenous glucagon and insulin in unstable diabetes. Diabetes 26:36-45, 1977.

[381] Rimoin, D. L.: Genetic syndromes associated with glucose intolerance. Ibid. p. 43-63.

[382] Rinderknecht, E., and Humbel, R.: The amino acid sequence of human insulin-like growth factor I and its structural homology with insulin. J. Biol. Chem. 1978, in press.

[383] Robertson, P. R., and Porte, D.: Andrenergic modulation of basal insulin secretion in man. Diabetes 22:1-8, 1973.

[384] Rosen, L. S., Fullerton, W. W., and Low, B. W.: Proinsulin: Further crystallization and x-ray diffraction studies of bovine and porcine prohormone. Arch. Biochem. Biophys. 152:569-73, 1972.

[385] Ross, B. D., Epstein, F. H., and Leaf, A.: Sodium reabsorption in the perfused rat kidney. Am. J. Physiol. 225:1165-71, 1973.

[386] Roth, J., Gorden, P., and Pastan, I.: "Big insulin": A new component of plasma insulin detected by immunoassay. Proc. Natl. Acad. Sci. USA 61:138-45, 1968.

[387] Rubenstein, A. H., Cho, S., and Steiner, D. F.: Evidence for proinsulin in human urine and serum. Lancet 1:1353-55, 1968.

[388] Rubenstein, A. H., Melani, F., Pilkis, S., and Steiner, D. F.: Proinsulin. Secretion, metabolism, immunological and biological properties. Postgrad. Med. J. 45 (suppl.) July 1969, 476-81.

[389] Rubenstein, A. H., Clark, J. L., Melani, F., and Steiner, D. F.: Secretion of proinsulin C-peptide by pancreatic beta cells

and its circulation in blood. Nature 224:697-99, 1969.

[390]Rubenstein, A. H.: The significance of immunoassayable insulin in urine. J.A.M.A. 209:254-56, 1969.

[391]Rubenstein, A. H., and Steiner, D. F.: Human proinsulin: some considerations in the development of a specific immunoassay. In Early Diabetes. Camerini-Davalos, R., and Levine, R., Eds. New York, Academic Press, 1970, pp. 159-69.

[392]Rubenstein, A. H., Welbourne, W. P., Mako, M., Melani, F., and Steiner, D. F.: Comparative immunology of bovine, porcine and human proinsulins and C-peptides. Diabetes 19:546-53, 1970.

[393]Rubenstein, A. H., Block, M. B., Starr, J., Melani, F., and Steiner, D. F.: Proinsulin and C-peptide in blood. Diabetes 21 (Suppl. 2):661-72, 1972.

[394]Rubenstein, A. H., Melani, F., and Steiner, D. F.: Circulating proinsulin: immunology, measurement, and biological activity. In Handbook of Physiology, Endocrinology I. Steiner, D. F., and Freinkel, N., Eds. Baltimore, Williams and Wilkins, 1972, pp. 515-28.

[395]Rubenstein, A. H., Pottienger, L. A., Mako, M., Getz, G. S., and Steiner, D. F.: The metabolism of proinsulin and insulin by the liver. J. Clin. Invest. 51:912-21, 1972.

[396]Rubenstein, A. H., Melani, F., and Steiner, D. F.: Proinsulin and C-peptide in human serum. In Methods in investigative and diagnostic endocrinology. Vol. 2B. "Peptide hormone." Berson, S. A., and Yalow, R. S., Eds. Amsterdam, North-Holland Publishing Co., 1973, p. 870-76.

[397]Rubenstein, A. H., Mako, M. E., and Horwitz, D. L.: Insulin and the kidney. Nephron 15:306, 1975.

[400]Rubenstein, A. H., Steiner, D. F., Horwitz, D. L., Mako, M. E., Block, M. B., Starr, J. I., Kuzuya, H., and Melani, F.: Clinical significance of circulating proinsulin and C-peptide. Rec. Progr. Horm. Res. 33:435-68, 1977.

[401]Salokangas, A., Smyth, D. G., Markussen, J., and Sundby, F.: Bovine proinsulin: amino acid sequence of the C-peptide isolated from pancreas. Eur. J. Biochem. 20:183-89, 1971.

[402]Samols, E., and Ryder, J. A.: Studies on tissue uptake of insulin in man using a differential immunoassay for endogenous and exogenous insulin. J. Clin. Invest. 40:2092-2102, 1961.

[404]Samols, E., and Turner, M. D.: Unpublished observations quoted In Carbohydrate Metabolism and its Disorders. Dickens, F., Randle, P. J., and Whelan, W. J., Eds. London, Academic Press, 1968, p. 298.

[405]Sandler, R., Horwitz, D. L., Rubenstein, A. H., and Kuzuya, H.: Hypoglycemia and endogenous hyperinsulinism complicating diabetes mellitus: application of the C-peptide assay to diagnosis and therapy. Am. J. Med. 59:730-36, 1975.

[406]Sando, H., Borg, J., and Steiner, D. F.: Studies on the secretion of newly synthesized proinsulin and insulin from isolated rat islets of Langerhans. J. Clin. Invest. 51:1476-85, 1972.

[407]Scarlett, J. A., Mako, M. E., Rubenstein, A. H., Blix, P. M., Goldman, J., Horwitz, D. L., Sternholm, M. R., and Olefsky, J. M.: Factitious hypoglycemia: diagnosis by measurement of serum C-peptide immunoreactivity and insulin-binding antibodies. Submitted for publication.

[408]Schatz, H., Otto, J., Hinz, M., Maier, V., Nierle, C., and Pfeiffer, E. F.: Gastrointestinal hormones and function of pancreatic islets: studies on insulin secretion, 3H leucine incorporation and intracellular free leucine pool in isolated pancreatic mouse islets. Endocrinology 94:248-56, 1974.

[409]Schlichtkrull, J., Munck, O., and Jersild, M.: The M-value, an index of blood-sugar control in diabetics. Acta Med. Scand. 177:95-102, 1965.

[410]Schlichtkrull, J.: Antigenicity of monocomponent insulins. Lancet 2:1260-61, 1974.

[411]Schlichtkrull, J., Brange, J., Christiansen, Aa. H., Hallund, O., Heding, L. G., Jørgensen, K. H., Rasmussen, S. M., Sørensen, E., and Vølund, Aa.: Monocomponent insulin and its clinical implications. Horm. Metab. Res. (Suppl.) Series No. 5:134-43, 1974.

[412]Schmidt, F. H.: Enzymatische Methode zur Bestimmung von Blut- und Harnzucker unter Berücksichtigung von Vergleichsuntersuchungen mit klassichen Methoden. Internist (Berlin) 4:554-59, 1963.

[413]Schmidt, F. H.: Methoden der Harn- und Blutzuckerbestimmung. In Handbook of Diabetes Mellitus. Vol. II. Pfeiffer, E. F., Ed. München, J. F. Lehmans, 1971, 913-46.

[415]Sells, R. A., Calne, R. Y., Hadjiyanakis, V., et al.: Glucose and insulin metabolism after pancreatic transplantation. Br. Med. J. 3:678-81, 1972.

[416]Seltzer, H. S., and Smith, W. C.: Plasma insulin activity: an index of insulogenic reserve in normal and diabetic man. Diabetes 8:417-24, 1959.

[417]Seltzer, H. S., Allen, W., Herron, A. L., and Brennan, M. T.: Insulin secretion in response to glycemic stimulus: relation of delayed initial release to carbohydrate intolerance in mild diabetes mellitus. J. Clin. Invest. 46:323-35, 1967.

[418]Service, F. J., Molnar, G. D., Rosevear, J. W., Ackerman, E., Gatewood, L. C., and Taylor, W. F.: Mean amplitude of glycemic excursions, a measure of diabetic instability. Diabetes 19:644-55, 1970.

[419]Service, F. J., Molnar, G. D., and Taylor, W. R.: Urine glucose analyses during continuous blood glucose monitoring. J.A.M.A. 222:294-98, 1972.

[420]Service, F. J., Rubenstein, A. H., and Horwitz, D. L.: C-peptide analysis in diagnosis of factitial hypoglycemia in an insulin-dependent diabetic. Mayo Clin. Proc. 50:697-701, 1975.

[421]Service, F. J., Horwitz, D. L., Rubenstein, A. H., Kuzuya, H., Mako, M. E., Reynolds, C., and Molnar, G. D.: C-peptide suppression test for insulinoma. J. Lab. Clin. Med. 90:180-86, 1977.

[422]Service, F. J., Go, V. L., Blix, P., and Nelson, R. L.: Studies of the direct effect of insulin on insulin, glucagon, GIP and gastrin secretion during maintenance of normoglycemia (euglycemic clamp.). Abstract 87. Diabetes 26 (Suppl. 1):374, 1977 (Abstr.).

[423]Sestoft, L., and Rehfeld, J. F.: Insulin and glucose metabolism in liver cirrhosis and in liver failure. Scand. J. Gastroenterol. (Suppl. 7):133-36, 1970.

[424]Shafrir, E.: Hyperlipidemia in diabetes. In Diabetes Mellitus. Sussman, K. E., and Metz, R. J. S., Eds. Vol. 4. American Diabetes Association, New York, 1975, pp. 221-28.

[425]Shen, S. W., Reaven, G. M., and Farquhar, J. W.: Comparison of impedence to insulin-mediated glucose uptake in normal subjects and in subjects with latent diabetes. J. Clin. Invest. 49:2151-60, 1970.

[427]Sherwin, R. S., Joshi, P., Hendler, R., et al.: Hyperglucagonaemia in Laennec's cirrhosis, the role of portalsystemic shunting. N. Engl. J. Med. 290:239-42, 1974.

[428]Sherwin, R. S., Hendler, R. J., and Felig, P.: Effect of diabetes mellitus and insulin on the turnover and metabolic response to ketones in man. Diabetes 25:776-84, 1976.

[429]Shima, K., Miyashita, T., and Tarui, S.: Urine CPR excretion in insulinoma. (Letter to the editor) J. Jap. Diabet. Soc. 20:270, 1977.

[430]Shima, K., Tanaka, R., Morishita, S., Tarui, S., Kumarhara, Y., and Nishikawa, M.: Studies on the etiology of "brittle diabetes." Diabetes 26:717-25, 1977.

[431]Shoemaker, W. C., Van Itallie, T. B., and Walker, W. F.: Measurement of hepatic glucose output and hepatic blood flow in response to glucagon. Am. J. Physiol. 196:315-18, 1959.

[432]Shurberg, J. L., Resnick, R. H., Koff, R. S., et al.: Serum lipids, insulin and glucagon after portacaval-shunt in cirrhosis. Gastroenterology 72:301-04, 1977.

[433]Sifferd, R. H., and du Vigneaud, V.: A new synthesis of carnosine, with some observations on the splitting of the benzyl group from carbobenzoxy derivatives and from benzylthio ethers. J. Biol. Chem. 108:753-61, 1935.

[434]Singal, D. P., and Blajchman, M. A.: Histocompatibility (HL-A) antigens, lymphocytotoxic antibodies and tissue antibodies in patients with diabetes mellitus. Diabetes 22:429-32, 1973.

[435]Siperstein, M. D., Unger, R. H., and Madison, L. L.: Studies of muscle, capillary basement-membrane in normal subjects, diabetic and prediabetic patients. J. Clin. Invest. 47:1973-99, 1968.

[436]Siperstein, M. D., Foster, D. W., Knowles, H. C., Levine, R., Madison, L. L., and Roth, J. J.: Control of blood glucose and diabetic vascular disease. N. Engl. J. Med. 296:1060-63, 1977.

[437]Snedecor, G. W., and Cochran, W. G.: Statistical Methods. 6th edit. Ames, Iowa State University Press, 1967.

[438]Snell, C. R., and Smyth, D. G.: Proinsulin: A proposed three-dimensional structure. J. Biol. Chem. 250:6291-95, 1975.

[439]Soeters, P., Weir, G., Ebeid, A. M., et al.: Insulin and glucagon following portacaval shunt. Gastroenterology 69:867, 1975.

[440]Soeters, P. B., and Fischer, J. E.: Insulin, glucagon, amino-acid imbalance, and hepatic encephalopathy. Lancet 2:880-82, 1976.

[441]Solomon, S. S., Brush, J. S., and Kitabchi, A. E.: Antilipolytic activity of insulin and proinsulin on ACTH and cyclic nucleotide-induced lipolysis in the isolated adipose cell of rat. Biochim. Biophys. Acta 218:167-69, 1970.

[443]Spitz, I. M., Rubenstein, A. H., Bersohn, I., Wright, A. D., and Lowy, C.: Urine insulin and renal disease. J. Lab. Clin. Med. 75:990-1005, 1970.

[444]Starr, J. I., and Rubenstein, A. H.: Metabolism of endogenous proinsulin and insulin in man. J. Clin. Endocrinol. Metab. 38:305-08, 1974.

[445]Starzl, T. E., Putnam, C. W., Chase, H. P., et al.: Portacaval shunt in hyperlipoproteinaemia. Lancet 2:940-44, 1973.

[446]Starzl, T. E., Porter, K. A., Kashiwagi, N., et al.: The effect of diabetes mellitus on portal blood hepatotrophic factors in dogs. Surg. Gynecol. Obstet. 140:549-62, 1975.

[447]Steiner, D. F., and Oyer, P. E.: The biosynthesis of insulin and a probable precursor of insulin by a human islet cell adenoma. Proc. Natl. Acad. Sci. USA 57:473-80, 1967.

[448]Steiner, D. F., and Clark, J. L.: The spontaneous reoxidation of reduced beef and rat proinsulins. Proc. Natl. Acad. Sci. USA 60:622-29, 1968.

[449]Steiner, D. F., Hallund, O., Rubenstein, A. H., Cho, S., and Bayliss, C.: Isolation and properties of proinsulin, intermediate forms, and other minor components from crystalline bovine insulin. Diabetes 17:725-36, 1968.

[450]Steiner, D. F.: Proinsulin and the biosynthesis of insulin. N. Engl. J. Med. 280:1106-13, 1969.

[451]Steiner, D. F., Clark, J. L., Nolan, C., Rubenstein, A. H., Margoliash, E., Aten, B., and Oyer, P. E.: Proinsulin and the biosynthesis of insulin. In Recent Progress in Hormone Research. Astwood, E. B., Ed. New York, Academic Press 1969, pp. 207-82.

[452]Steiner, D. F., Clark, J. L., Nolan, C., Rubenstein, A. H., Margoliash, E., Melani, F., and Oyer, P. E.: The biosynthesis of insulin and some speculations regarding the pathogenesis of human diabetes. In The Pathogenesis of Diabetes Mellitus. Proceedings of the Thirteenth Nobel Symposium. Cerasi, E., and Luft, R., Eds. Stockholm, Almqvist and Wiksell, 1970, pp. 123-32.

[453]Steiner, D. F.: Cocrystallization of proinsulin and insulin. Nature 243:528-30, 1973.

[454]Steiner, D. F., Terris, S., Emdin, S. O., Peterson, J. D., and Falkmer, S.: Evolution and comparative biology of islet secretory products. In Early Diabetes in Early Life. Camerini-Davalos, R., Ed. New York, Academic Press, 1975, pp. 41-48.

[455]Steiner, D. F.: Peptide hormone precursors: Biosynthesis, processing and significance. In Peptide Hormones. Parsons, J. A., Ed. London, Macmillan Press, 1976, pp. 49-65.

[456]Steiner, D. F.: Insulin today. Diabetes 26:322-40, 1977.

[457]Sternberger, L. A., Hindy, P. H., Cuculis, J. J., and Meyer, H. G.: The unlabeled antibody enzyme method of immunochemistry. Preparation and properties of soluble antigen-antibody complex (horseradish peroxidase-antihorseradish peroxidase) and its use in identification of spirochetes. J. Histochem. Cytochem. 18:315-33, 1970.

[458]Stevenson, R. W., Parsons, J. A., and Alberti, K. G. M. M.: Insulin infusion into the portal and peripheral circulations of unanaesthetized dogs. Clin. Endocrinol., in press.

[460]Stoll, R. W., Touber, J. L., Menahan, L. A., and Williams, R. H.: Clearance of porcine insulin, proinsulin, and connecting peptide by the isolated rat liver. Proc. Soc. Exp. Biol. Med. 133:894-96, 1970.

[461]Storey, H. T., Beacham, J., Cernosek, S. F., Finn, F. M., Yanaihara, N., and Hofmann, K.: Studies on polypeptides. II. Application of S-ethylcarbamoylcysteine to the synthesis of a protected heptatetracontapeptide related to the primary sequence of ribonuclease T₁. J. Am. Chem. Soc. 94:6170, 1972.

[462]Strober, W., and Waldmann, T. A.: The role of the kidney in the metabolism of plasma proteins. Nephron 13:35-66, 1974.

[463]Sundler, F., Alumets, J., Holst, J., Larsson, L. I., and Håkanson, R.: Ultrastructural identification of cells storing pancreatic-type glucagon in dog stomach. Histochemie 50:33-37, 1976.

[464]Sönksen, P. H., Tompkins, C. V., Srivastava, M. C., and Nabarro, J. D. N.: A comparative study on the metabolism of human insulin and porcine proinsulin in man. Clin. Sci. Molecul. Med. 45:633-54, 1973.

[465]Tager, H. S., and Steiner, D. F.: Primary structures of the proinsulin connecting peptides of the rat and the horse. J. Biol. Chem. 247:7936-40, 1972.

[466]Tager, H. S., Emdin, S. O., Clark, J. L., and Steiner, D. F.: Studies on the conversion of proinsulin to insulin. II. Evidence for a chymotrypsin-like cleavage in the connecting peptide region of insulin precursors in the rat. J. Biol. Chem. 248:3476-82, 1973.

[467]Tanimura, T., Pisano, J. J., Ito, Y., and Bowman, R. L.: Droplet countercurrent chromatography. Science 169:54-56, 1970.

[468]Tattersall, R., and Pyke, D. A.: Diabetes in identical twins. Lancet 2:1120-25, 1972.

[469]Tattersall, R.: The inheritance of maturity-onset type diabetes in young people. In The Genetics of Diabetes Mellitus. Creutzfeldt, W., Köbberling, J., and Neel, J. V., Eds. Springer-Verlag, Berlin, Heidelberg, New York, 1976, pp. 88-94.

[470]Technicon AutoAnalyzer II, clinical method No. 02, March, 1972.

[471]Terris, S., and Steiner, D. F.: Binding and degradation of ^{125}I-insulin by rat hepatocytes. J. Biol. Chem. 250:8389-98, 1975.

[472]Terris, S., and Steiner, D. F.: Retention and degradation of ^{125}I-insulin by perfused rat livers. J. Clin. Invest. 57:885-96, 1976.

[473]Thomas, S. K., Taylor, W. F., and Molnar, G. D.: Continuous blood glucose analysis in ambulatory fed subjects. III. Nitrogen balance and urinary ketone bodies related to other measurements in the characterization of unstable diabetes. Mayo Clin. Proc. 49:28-33, 1974.

[474]Thomsen, M., Platz, P., Andersen, O. O., Christy, M., Lyngsøe, J., Nerup, J., Rasmussen, K., Ryder, L. P., Staub Nielsen, L., and Svejgaard, A.: MLC typing in juvenile diabetes mellitus and idiopathic Addison's disease. Transplant Rev. 22:125-47, 1975.

[475]Toccafondi, R., Rotella, C. M., Tanini, A., and Arcongeli, P.: Plasma proinsulin-like components and insulin in chronic liver disease. Horm. Metab. Res. 9:101-05, 1977.

[476]Tsalikian, E., Dunphy, T. W., Bohannon, N. V., Lorenzi, M., Gerich, J. E., Forsham, P. H., Kane, J. P., and Karam, J. H.: The effect of chronic oral antidiabetic therapy on insulin and glucagon responses to a meal. Diabetes 26:314-21, 1977.

[477]Turner, R. C., Oakley, N. W., and Nabarro, J. D. N.: Changes in plasma insulin during ethanol-induced hypoglycaemia. Metabolism 22:111-21, 1973.

[478]Turner, R. C., and Johnson, P. C.: Suppression of insulin release by fish-insulin induced hypoglycaemia: with reference to the diagnosis of insulinomas. Lancet 1:1483-85, 1973.

[479]Turner, R. C., and Harris, E.: Diagnosis of insulinomas by suppression tests. Lancet 1:188-90, 1974.

[480]Turner, R. C., Harris, E., and Bloom, S.: Two disorders of deficient glucose-induced insulin secretion: a "quantitative" defect (low V max) in latent diabetes, and decreased sensitivity of beta cells in women who produced LFD babies. 11th Annu. Meet. Eur. Assoc. Study Diabetes, Munich. Diabetologia 11:380, 1975 (Abstr.).

[481]Turner, R. C., McCarthy, S. T., Holman, R. R., and Harris, E.: Beta cell function improved by supplementing basal insulin secretion in mild diabetes. Br. Med. J. 1:1252-54, 1976.

[482]Turner, R. C., and Holman, R. R.: Insulin rather than glucose homeostasis in the pathophysiology of diabetes. Lancet 1:1272-74, 1976.

[483]Turner, R. C., and Holman, R.: Pathophysiology of diabetes. Lancet 2:856, 1976.

[484]Turner, R. C., Mann, J. I., Simpson, R. D., Harris, E., and Maxwell, R.: Fasting hyperglycaemia and relatively unimpaired meal responses in mild diabetes. Clin. Endocrinol. 6:253-64, 1977.

[485]Turner, R. C., Harris, E., Bloom, S. R., and Uren, C.: Relation of fasting plasma glucose concentration to plasma insulin and glucagon concentrations. Studies in latent diabetics and women who have produced large-for-dates babies. Diabetes 20:166-71, 1977.

[486]Turner, R. C., Holman, R., and Hockaday, T. D. R.: Feedback analysis of beta cell deficit and insulin resistance in diabetes. Submitted for publication in Diabetes, 1977.

[487]Turner, R. C., Holman, R. R., and Eaton, P.: Beta cell recovery with regular meal/insulin infusion regime in acute, ketotic diabetes. Submitted for publication.

[489]Unger, R. H.: Diabetes and the alpha cell. Banting Memorial Lecture 1975. Diabetes 25:136-51, 1976.

[490]University Group Diabetes Program: A study of the effects of hypoglycemic agents on vascular complications in patients with adult-onset diabetes. Diabetes 19 (Suppl. 2):747-830, 1970.

[491]University Group Diabetes Program: A study of the effects of hypoglycemic agents of vascular complications in patients with adult-onset diabetes. VI. Supplementary report on nonfatal events in patients treated with tolbutamide. Diabetes 25:1129-53, 1976.

[492]Vallance-Owen, J., Hurlock, B., and Please, N. W.: Plasma insulin activity in diabetes mellitus measured by the rat diaphragm technique. Lancet 2:583-87, 1955.

[493]Varandani, P. T.: A convenient preparation of reduced and S-sulfonated A and B chains of insulin. Biochim. Biophys. Acta 127:246-48, 1966.

[494]Waddell, W. R., and Sussman, K. E.: Plasma insulin after diversion of portal and pancreatic venous blood to vena cava. J. Appl. Physiol. 22:808-12, 1967.

[495]Walker, C., Peterson, W., and Unger, R.: Blood ammonia levels in advanced cirrhosis during therapeutic elevation of the insulin:glucagon ratio. N. Engl. J. Med. 291:168-71, 1974.

[496]Weber, B., and Neumann, B.: Insulin response to glucagon in diabetic children. Z. Kinderheilkd. 115:175-85, 1973.

[497]Wecklicki, W. B., Soeldner, J. S., and Cohn, L. H.: Portacaval diversions for severe hypercholesterolaemia: report of a case with measurements of glucose tolerance, insulin and glucagon levels. Arch. Surg. 112:634-36, 1977.

[498]Weichselbaum, A.: Über die Veränderungen des Pankreas bei Diabetes mellitus. S. uber. Wien. Akad. - Math. - Naturw. Kl. 119 Abt. III, 73-281, 1910.

[499]Werner, W., Rey, H. G., and Wielinger, H.: Über die Eigenschaften eines neuen Chromogens für die Blutzuckerbestimmung nach der GOD/POD-Methode. Z. Anal. Chem. 252:224-28, 1970.

[500]West, K. M.: Diet therapy of diabetes: an analysis of failure. Ann. Intern. Med. 79:425-34, 1973.

[501]Wilkerson, H. L. C.: Diagnosis, oral glucose tolerance tests. In Diabetes Mellitus. American Diabetes Association, New York, 1964, pp. 31-34.

[502]Williamson, J. R., and Kilo, C.: Current status of capillary basement-membrane disease in diabetes mellitus. Diabetes 26:65-73, 1977.

[503]Wollmer, A., Brandenburg, D., Vogt, H. P., and Schermutzki, W.: Reduction/reoxidation studies with cross-linked insulin derivatives. Hoppe-Seylers Z. Physiol. Chem. 355:1471-76, 1974.

[504]Woods, K. R., and Wang, K.-T.: Separation of dansyl amino acid by polyamide layer chromatography. Biochim. Biophys. Acta 133:369-70, 1967.

[505]Yalow, R. S., and Berson, S.: Immunoassay of endogenous plasma insulin in man. J. Clin. Invest. 39:1157-75, 1960.

[506]Yalow, R. S., Glick, S. M., Roth, J., and Berson, S. A.: Plasma insulin and growth hormone levels in obesity and diabetes. Ann. N.Y. Acad. Sci. 131:357-73, 1965.

[507]Yanaihara, N., Yanaihara, C., Dupuis, G., Beacham, J., Camble, R., and Hofmann, K.: Studies on polypeptides. XLII. Synthesis and characterization of seven fragments spanning the entire sequence of ribonuclease T₁. J. Am. Chem. Soc. 91:2184, 1969.

[508]Yanaihara, N., Hashimoto, T., Yanaihara, C., and Sakura, N.: Studies on the synthesis of proinsulin. I. Synthesis of partially protected tritriacontapeptide related to the connecting peptide fragment of porcine proinsulin. Chem. Pharm. Bull. 18:417, 1970.

[509]Yanaihara, N., Yanaihara, C., Hashimoto, T., Sakagami, M., and Sakura, N.: On proinsulin synthesis. In Chemistry and Biology of Peptides. Meienhoger, J., Ed. Ann Arbor, Science Publishers, 1972, p. 287.

[510]Yanaihara, N., Yanaihara, C., Hashimoto, T., Sakura, N., and Sakagami, M.: Proceedings of the 9th symposium on peptide chemistry. Yanaihara, N., Ed. Osaka, Protein Research Foundation Press, 1972, p. 122.

[511]Yanaihara, N., Sakura, N., Yanaihara, C., and Hashimoto, T.: Studies on the synthesis of proinsulin. III. Synthesis of polypeptides related to the connecting peptide segment of bovine proinsulin. J. Am. Chem. Soc. 94:8243-44, 1972.

[512]Yanaihara, N., Hashimoto, T., Yanaihara, C., Sakagami, M., and Sakura, N.: Synthesis of polypeptides related to porcine proinsulin. Diabetes 21 (Suppl. 2):476-85, 1972.

[513]Yanaihara, N., Hashimoto, T., Yanaihara, C., Sakagami, M., Steiner, D. F., and Rubenstein, A. H.: Synthesis of human connecting peptide derivatives and their immunological properties. Biochem. Biophys. Res. Commun. 59:1124-30, 1974.

[514]Yanaihara, N., Sakura, N., Yanaihara, C., Hashimoto, T., Rubenstein, A. H., and Steiner, D. F.: Synthesis and immunological evaluation of bovine proinsulin C-peptide analogues. Nature 258:365-66, 1975.

[515]Yanaihara, N., Hashimoto, T., Yanaihara, C., Sakura, N., and Sakagami, M.: Chemistry and immunochemistry of prohormone. In Endocrine Gut and Pancreas. Fujita, T., Ed. Amsterdam, Elsevier Scientific Publishing Co., 1976, p. 186.

[516]Yanaihara, N., Sakagami, M., Sakura, N., Iizuka, Y., Nishida, T., Hashimoto, T., and Yanaihara, C.: Synthetic C-peptides and proinsulin synthesis. Proc. 9th Congr. IDF, New Delhi, 1976, in press.

[517]Yu, N.-T., Liu, C. S., and O'Shea, D. C.: Laser Raman spectroscopy and the conformation of insulin and proinsulin. J. Molec. Biol. 70:117-32, 1972.

[518]Yu, S. S., and Katabchi, A. E.: Biological activity of proinsulin and related polypeptides in the fat tissue. J. Biol. Chem. 248:3753-61, 1973.

[519]Ziegler, M., Hahn, H. J., and Klatt, D.: Influence of isolated insulin antibodies on the insulin secretion of the islets of Langerhans in vitro. Diabetologia 8:148-49, 1972.

[520]Zilker, Th., Kränzlin, Th., Schweigart, U., Ermler, R., and Bottermann, P.: Radioimmunologische Seruminsulinbestimmung beim, manifesten Diabetes mellitus als Parameter der Therapieplanung. Med. Klin. 71:761-67, 1976.

[521]Zilker, Th., Wiesinger, H., Ermler, R., Schweigart, U., and Botterman, P.: C-Peptid-Konzentration im Serum in Abhängigkeit von der Nierenfunktion. Klin. Wochenschr. 55:471-74, 1977.

[522]Zühlke, H., Steiner, D. F., Lernmark, A., and Lipsey, C.: Carboxypeptidase B-like and trypsin-like activities in isolated rat pancreatic islets. In Polypeptide Hormones: Molecular and Cellular Aspects. CIBA Foundation Symposium 41. Elsevier/Excerpta Medica, Amsterdam, 1976, pp. 183-95.

[523]Buckle, A. L. J., Nattrass, M. Cluett, B. E., Stubbs, W. A., Walton, R. J., Alberti, K. G. M. M., and Clemens, A. H: Blood metabolite concentrations in diabetics: effect of normalisation of blood glucose using a glucose controlled insulin infusion system (GCIIS). 13th Annu. Meet. Eur. Assoc. Study Diabetes, Geneva. Diabetologia 13:385 Abstr. 1977.

[524]Mirouze, J., Selam, J. L., Pham, T. C., Mendoza, E., and Orsetti, A.: Remission of juvenile diabetes with severe insulin deficiency induced by early transient insulin therapy assessed by the artificial beta cell. 13th Annu. Meet. Eur. Assoc. Study Diabetes, Geneva. Diabetologia 13:419 Abstr. 1977.

[525]Turner, R. C., Holman, R. R., and Harris, E.: Beta cell recovery with a regular meal/insulin infusion regime in new, severe, ketotic diabetics. 13th Annu. Meet. Eur. Assoc. Study Diabetes, Geneva. Diabetologia 13:437 Abstr. 1977.